Deep Keel

By Joel J Guttormsen

DEEP KEEL MINER
By (JJG)

June 2024

Dedicated to:

"To my beloved wife and eternal companion!"

Also, I reserve this for my furry companion, a dog, who was stuck with me abroad during COVID-19.

"To Jemma the Golden, friend, companion, guard, support, and a fine listener."

We both made it home!

Sorry, she was so young when she passed, but I will see her again!

Table of Contents

Chapter 1
Cirrus System

Two hundred fifty million kilometers from the planet Cirrus -Inner Asteroid belt 2nd planetary position Mining Station.

The Cirrus System is one of the most important solar systems other than ancient Earth. The system is the primary source of two essential commodities used in interstellar travel by the expanding Human Universe. The system is home to a class 2F sun, a garden-like Planet Cirrus Major (Planet Five), and an expansive asteroid field controlled by the Guild of Free Miners Cirrus Minor. The Guild provides services to the miners at reasonable costs and negotiates sales with outside groups such as the mega-Earth corporation Mantle Core through its subsidiary Mantle Works. While most of the guild's contracts are with the Mantle Works, no love is lost between the two groups. They have had a long history of minor skirmishes in the asteroid belt. The Guild headquarters is a significant mining base in the Asteroids. The mining base is an organized, accessible community of families and friends bonded together by a common goal of mining. Most families have been in the asteroid mining business for generations. Everyone is trained to be a miner, and education is the top priority. Miners are highly skilled scientists who like to cut rocks and dig out their treasures while paid very handsomely.

John Erricsen stood at the beckoning call of his galactic history teacher. The class assignment was to provide a historical synopsis of one of the elements of human expansion. He had put together a five-minute talk and a holographic show of the various components relevant to his project. In addition, a technical document of three hundred-plus pages was required. The technical discussion alone was worth 40% of the class grade. He had to pass to graduate from the mining base school! He had to graduate to drive the mining sled in his family-run business and continue his education. He loved science but hated history. John's paper focused on the history of Draconia mining, including the technical specifications and operation of the equipment used to mine the material. It read like an operations manual. He had to be careful not to show his natural distaste for the Mantle Works Corporation, which provided staff at his school. John had implied negative feelings about the Mantle Works Corporation that once severely affected his grades, including a misconduct letter in his personnel folder. This event potentially damaged his opportunity for further education. The Mantle Work Corporation was rugged if you owed them a debt. Fortunately, the station was run by the Mining Guild. The guild was law in the asteroids but was tied to the Mantle Works Corporation by ninety percent of its contracts.

Erricsen boldly began: "The Cirrus System lies 34 au's from the Earth's core solar system. The development of ion drive technology fueled the human expansion era from 3510 A.D. and opened a series of wormholes, which allowed the discovery of the Cirrus system. Humanity's galactic growth was fueled by finding class M systems containing Nickel-Iron cored planets due to the human thirst for resources to build the emerging star empire. One of the critical resources was the catalyst used in the

Ion drive system known as Draconia, which was discovered on Earth in 2980 by deep core miners seeking large diamond masses used as a critical component in force field generation.

Originally, Draconia was investigated and classified as an extremely rare gemstone due to its mesmerizing iridescent qualities that changed with varying light and magnetic fields. Draconia would eventually be a critical catalytic component of Ion Drives. Since the stones were scarce and only available through deep Earth mining sources, they were highly prized. These "deep mining" companies were known as Mantle Excavators, Siberian Deep, and Point Source. They became mega-conglomerates by expanding from deep-core mining into asteroid mining.

At this point in history, the nuclear engine was developed and used extensively for asteroid mining. However, the gravities experienced by the accelerations produced by this engine could only be managed by the gravity force field envelopes driven through diamond masses. The "deep miners" had tight control of the critical resources and terminated competition (legally or by other means). They eventually merged into a single entity known as Mantle Works. With the discovery of Ion drive, utilizing the newly discovered, exotically rare Draconia, and under the supervision of Mantle Works, the company's future was secured. Like their earliest diamond counterparts, they established iron-handed control of the commodity markets required for the expansion." His teacher coughed at the last sentence. John quickly corrected, "Sorry, the company devised the means to maintain its competitive edge." He promptly corrected himself, understanding he had nearly crossed the line.

John's twin sister, Tanya, rolled her eyes, thinking, 'There he goes again, trying to get exported to the labor camps at the Mantle Works ore processing center.'

John continued with a 3-D projection of a bipyramidal hexagonal rendition of a Draconia crystal full of swirling iridescent colors. "This, he explained, is the goal of independent miners - Draconia. Small amounts of less than 10 milligrams are required to fuel the catalytic reaction of the Ion drive. The gemstone is currently sold for 10 million credits per gram. It is so rare that only 25,500 grams have ever been found. Deep core mining on Earth was the only source of this material until a recent discovery in the Cirrus asteroid belt. That load was only reported to be about 5 grams; however, the actual amount has never been published. The miner disappeared just after the discovery was announced. Mantle Works reported the miner living happily on the beach in Southern California."

John switched slides to a holograph of the Cirrus Asteroid belt. "Our asteroid belt had the characteristics of what scientists describe as a class M planet. Recreations of the system suggest that 1000 to 5000 years ago, this planetary body had a catastrophic event that tore the planet apart. The event was not simple, however". He changed the projection to a graphic 3-D image of the planet before it was destroyed. The hologram flashed an intense white light followed by a tremendous ear-shattering bang. The teacher, leaning against his desk, lost his footing and dropped like a rock to the floor. Most of the class jumped under their desks where emergency survival bubbles were stored. John's sister calmly sat through the commotion and said in the calm after the storm, "Monkey," her nickname for her brother, "that may have just been the dumbest special effect ever pulled by you, ever!"

John looked at his teacher with a reddened face, picking himself up from the floor. "Mister Erricsen, I think you just won the prize for the best attention-grabbing stunt for this project!"

The door burst open, and two amour-suited security guards burst through the door. "I am sorry," John said while raising his hands, "that was not supposed to happen."

One of the guards looked at Mister Grotingen, the teacher, and asked, "Randy, everything ok?"

The teacher smiled slyly, "Well, Jim, it looks like we blew up a planet and started an impromptu depressurization drill at the school. I could not have planned it better. Anyway, the students were due for a drill."

Then he looked at Tanya and said, "Miss Erricsen, we need to complete the drill by having you on your knees under the desk with your head down and hands over your head. Please assume the position."

He turned to John. "I will assume you are a casualty and were fictitiously blown out into space."

Tanya said, smiling and moving under her desk," Thanks, Mister Grotingen, that would remove a royal pain from my side."

Mister Grotingen cleared his throat. "Class, now that you are awake, please sit in your chairs. Jim looks like we are under control. Stick around if you wish- John was reviewing 100 million years of history. The next twenty might get interesting". The class sat back down with some joviality. The teacher looked at John, whose face was returning to a more normal color. "John, you now have seven minutes of remaining time!"

John took a deep breath and began. "As you are aware, the Cirrus system has several planets. Only two were originally in the Goldilocks zone. Cirrus Major lies exterior to Cirrus Minor. It is rare to see an asteroid belt inside the Goldilocks zone as most debris belts are located just outside this zone. Nickel Iron core planets are generally developed in the interior portions of a solar system. Original explorers were focused on the class M Cirrus Major and established the colony a little over 500 years ago. Due to its molten Ni-Fe iron-rich core, Cirrus Major was rich in near-surface precious metals and rapidly became a mining colony. Mantle Works began deep core mining targeting the massive diamond clusters needed for force fields two hundred and fifty years ago. Like Earth, they were successful and even found a little Draconia, but they expected more." John continued through his presentation.

"Also, like Earth, the Draconia occurred near the diamond masses. Draconia is formed as Ni-Fe-telluric-carbon. Geologists noticed the massive diamond clusters in second bench intrusives along mantle keels developed under thick continental crust. The second bench is due to deeper magma injection from crustal subduction-related melting. These magmas are composed of eclogite and peridotite, which carry micro diamonds, the seeds for the diamond clusters. Mining operations explored and mined the mantle keels, where these features were identified through surface kimberlites. These were identified by tracing the kimberlite surface deposits to their 120 to 140 km sources in the deep Mantle mining process. The first-generation diamond shields originally developed by Hall Klinker, working with the Deep Mining Company, were the technology used to make these zones survivable. The Klinker shields created a habitable bubble for miners to control extraordinary pressure and manipulate temperature-resistant mining robots. The mining geologist built incredibly accurate models with deep earthquake-generated shear wave velocity analysis.

John updated his presentation. The Cirrus System now had twelve planets instead of eleven and an asteroid field. The two blue planets with brown continents stood out within the solar system. Labels hung on the worlds that included Cirrus Major and Cirrus Minor.

John continued, "Cirrus Minor was another iron core world that contained an oxygen atmosphere until its destruction. We know this because significant sedimentary basin fragments have been found in the asteroid belt. The fragments contain sandstone, shale, and carbonates consistent with those in other class m systems. Residual water extracted in these sediments is consistent with sea salt water containing roughly 35,000 parts per million chlorides."

He changed the hologram to show the same system where the planet existed now lay a massive asteroid field. The sedimentary asteroids' color changed to blue. "The nickel core was a clear target for mining as the asteroids are metallic, dense, and contain a ready abundance of base metals such as iron, titanium, lead, and zinc." The holographic projector changed the asteroid's color to metallic silver, with hues toned by the dominant metal. The rest of the asteroids comprise the upper, transition, and upper-lower mantle. It is in this area that the first diamond clusters were found. Draconia has been a more difficult nut to crack until a recent discovery by Potter Johns five years ago. Potter claimed to have cracked the code. He soon disappeared, and this has developed into a treasure seeker's legend. That covers about 4.5 billion years of history with a few minor stops on the way."

"Thank you, John," his teacher took the reins back. "I don't know if we just talked about diamond history or a history lesson in planetary genetics." The class laughed, which lightened the mood considerably. Mr. Grotingen continued, "It was above average, including the depressurization drill in the middle. Great holographic projection. I will dispense with questions except one." He theatrically paused. "John, the graphics were so good a normal family computer could not have driven them. Can you tell me what you used?" Mister Grotingen expected to hear that John had gotten some time on the school's nebula cluster computing array as he wanted others to use the system more. John broke into a big grin.

His sister groaned, and John's best friend, Tony, blurted out, "Mister Grotingen, you have never heard of John's *super-bomb* computer?" The teacher shook his head, "No, Mister Tony, I have not, but we will let John answer, which is much more polite."

As the students referred to him, John could not determine if Groni was truly kind-hearted or if he thought all miner children were impolite animals. He, however, could not duck the question. He had to be careful with his answer. "Mister Grotingen, I have built a diamond lithium state-of-the-art computer from scraps I found in the junkyard. I special ordered five Johnson 5400 processors from Cirrus Major with some inheritance money. It is helium-cooled and clocks 2 million zetta-flops (floating-point operations per second). It's grease lightning fast with four 500 graphics modules. It's faster than all the computers in this school, even if they were clustered."

"Sounds like you need a license for speed!" the teacher replied. After finding some abandoned military-grade computers from a fighter in the dump pile, John did not tell the class that he had 12 processors. He had also recently installed a damaged AI chip with the letters JEMMA stamped across it. He had found this in the scrapped test ship but did not fully understand its function other than he had doubled his processing speed. He was installing it in the family's old mining sled that he, Tony, his sister, and probably one of their mothers would be mining with during the summer break before university.

The bell rang. Mr. Grotingen announced, "Kyle, Manny T., and Rebecca Wolf will do their final presentation next Monday."

"Thanks for the show." The security guard, Jim, and his companion led the charge out of the room. The youth were ready for free time, and security had duties.

Tony met John and Tanya as they stepped outside the classroom. "Man, John, you put that on steroids," He continued. "Watching Cirrus Minor explode was epic. Mister Jim does not visit many classrooms yearly, which is a win. And Mister Grotingen's reaction was softhearted!"

Tanya said, "Yeah, that on the weird meter was about a seven. I thought he was going to hand you your butt."

John nodded his head in agreement. "Trust me, the blinding light and explosion were not as I had run them last night. Tanya, did you modify the file after I crashed last night?"

Tanya, slightly in front of the young men, stopped in her tracks, turned around, and went nose-to-nose with John. "While I enjoy a good prank, I would never sabotage your project. We need another pilot for mining once we graduate in a few weeks unless you want Mom," She said factually.

Tony piped in as she turned back, and they continued to walk down the mining station homeward. "Must have been that gremlin the miners talk a bunch about."

Tanya interjected, "Come on, Tony, that is just a wise tail. Dad probably saved it and shut your system down last night. He is not the best on new computers and may have hit an unintentional button. You did leave it on, and you know how he is about station electricity. You know what Dad always says: "Shut down to keep Alli poor." She was, of course, referring to Allison Jiggs, the iron-fisted CEO of Mantle Works."

Tony's face turned a little grey. "Tanya, you have to be careful. You know how these halls have ears. Just using her name can bring unwanted company."

John ignored him. "Yeah, Dad did turn it off. He still does not know about the little Californium source I stuffed in the box. It rocks and will last a long time."

Tanya looked quizzically at John. "Why did you not tell Mister Grotingen about what you have on your computer? He would have been impressed. I was surprised you stopped short with that little ego of yours."

John looked at her. "It is called military grade. Based on a run I made last night, that beast in that computer box may be the fastest on the station. If he knew, he might pass it up the line. And Jim was in the room. He is cool, but his buddy is a brown-nosing ladder climber and would have run to tell on me. I could not risk it.

Tony looked surprised. "One of these days, they will realize where they sent the experimental craft that they crashed."

"Yes, but I hope they will think they stripped all the valuable salvage before dumping it. I helped them with that, but probable weapon system components remain." John answered.

Tanya said, "So you are about to put your computer in charge of our mining sled! Is it safe?"

John was surprised. "I only pillaged their main core computer components that were not damaged. I did not grab anything except one chip that I did not recognize. That one chip had a single name on it and some minor scratches. I hooked it up on an ancillary system a couple of days ago. I could see it nearly doubled the system's speed, but I could not figure out how because it was on an auxiliary board. It is safe as a gravity sled, and I will put my life on it".

Gravity sleds were platforms where miners carried tools, plasma cutters, a mini-lab, and mined ore. They were as fast as molasses on a subzero morning when your breath freezes in the air. These invaluable sleds freed the Astro-miners from carrying heavy tools on their pressure suits. The sled also doubled as an emergency shelter.

"Hey, you guys going to the senior dance tomorrow - night?" Tony changed the subject.

Tanya perked up, "Yup, Rob Bikner asked me to go a couple of weeks ago. Mom's been working on just the right outfit. We have been having fun. Since John told Mom he was not going, she arranged a date with Sue 'Jaws' Thompson. He has to go to the dance, and we are having dinner on a double date. It should be more fun seeing John squirm all night, knowing Sue might try to kiss him."

Tony reacted, "Sue's the school's candidate for the highest iron in her blood. I thought she was a miner's dream with a mouth full of metal". Tony laughed at his old joke.

Sue was challenged with crooked teeth and poor care as a child. Her only option to correct the issues was the ancient technique of using wires on the teeth, known as braces. Hers were silver, causing her to be shy with her smiles.

John looked quizzically at Tony. "The wrong kind of metal, and she is a nice person. She works hard and is quiet."

"She is one of my best friends, Tony!" Tanya added.

Tony continued his verbal assault. "I think I heard a tinny clink-clink in the lunchroom as she was tearing a piece of vanadium-steel-plate flank steak the other day. Then, after lunch, I was blinded by the light reflecting off a light when she smiled at me. That sent shivers up my back. Just think, John, what will happen to you if you upset her".

John rolled his eyes. "Tony lay off of her. She is nice. Just give her a chance. Who knows, maybe you will marry her one day!"

They stopped at Tony's door, and Tony replied. "Just space me now. You guys stay out of trouble tonight, and I will see you tomorrow. John, try some metal shine on your teeth." He stopped halfway into his family's mining quarters.

Tanya smacked Tony in the arm. "Tony, remember what happens when you live in a plasto-steel house. What comes around goes around! Come on, John, Mom is waiting for me. You know, dress fitting and girl stuff."

They walked down the station's spine to the next door while Tony disappeared through his. John entered the code, but Tanya elbowed him and got through the door first. "Hey, mom, we are home!"

"OK, Tanya, I hope you can try this on. I have worked on it most of the day. If we have to alter it, I want as much time as possible," Their mom answered from the living room. "John, get your homework done before you work on sled Two. Your dad has built a tight schedule to prepare for the summer mining season."

"OK, Mom," John headed to his room. His room was a jumble of computer parts with a large curved screen and a blue box with six high-pressure hoses connected to a helium tank. His ceiling was a montage of the mining station set against a background of the Cirrus Minor Asteroid belt. It was striking.

John stepped into his room and said, "Computer on." Nothing happened.

"Computer on!" Again, nothing happened. Puzzled, he walked over to his desk, stepping across a box of wire bridals. He touched his keyboard, which immediately registered him, and his plasma screen jumped alive with a message:

'Upgrading quantum board — refiguring graphic interfaces - modifying system processing to optimum efficiency…… John, I am ready - I have a NAME - call me JEMMA, please."

John picked his jaw up and stuttered. "What the Hell?"

Chapter 2
Mantle Works - Homeworld

Four hundred years ago, Mantle Works moved its headquarters to Cirrus Major after continuous conflicts with Earth's new global government. The people of Earth finally established a worldwide united government and wanted the asteroid miners operating in the Earth's Solar System to pay taxes to the Earth. Mantle Works was caught between the tax-hungry government and the asteroid miners supplying Earth's new expansion into the Galaxy. That was the good news. The new global governments aimed to control all businesses and tightly control the expanded population stimulated by peace. Mantle Core, however, captured and managed the global economy with its Draconian Ion Drive, which they shamelessly defended. They were targeted with global hate cancel culture because they threatened the Earth bureaucrats. Gold was the old trade standard replaced by diamonds and, more recently, Draconia. Mantle works controlled most deep mantle mining operations and ninety-eight point five percent of the existing Draconia. Mantle Works escaped Earth in a cloud of shadow operations, and Mantle Works built eight space destroyers on Mars. Earth forces comprising two destroyers and a light cruiser pursued the last Mantle Core departures from Earth. A small battle ensued, destroying the Mars spaceport in an aggressive surprise attack by the Earth Fleet. Two technologically advanced Mantle Works destroyers visited the Earth Forces, celebrating their surprise victory and destroying the Cruiser. They also critically maimed one of the destroyers and left the other venting air. Earth did not know Mantle Works had shipped a massive city to the Cirrus System via their enormous fleet of ore carriers equipped with Ion engines. That was twelve generations ago. Mantle Works kept its dominance in the Earth's asteroid belt with its fleet of advanced destroyers. During the past one hundred years, they exported their older versions of space technology, such as generation one and two destroyer designs. That allowed Mantle Core to sell ion engines to Earth because they controlled the flow of a minimal amount of draconia, the crucial catalyst in the ion chamber reaction start-up.

The Mantle Works spire sat in the center of the sprawling megalopolis of Cirrus Five. The city was developed along the ancient lake terraces of Johns Lake. The lush jungle terraces were a striking contrast to the tall mirror reflections of the city's ferrous steel skyscrapers. The Mantle Works building was the tallest, sitting on the highest bench. Sitting at the top of the Mantle Works building was the expansive office of the president and CEO of the company, Allison Jiggs. In her office, sitting on a glass pedestal, sat a small, delicate hexagonal crystal that irradiated the room with iridescent light flashes as it slowly rotated on its stand. Allison sat mesmerized by the crystal of Draconia that had initially been mined in the deep core mining on Earth. This specimen had a long, violent history where most of the previous owners had met unpredictable violent deaths. Her ownership of the Crystal was no exception; the previous owner had taken two weeks to provide the vault codes to access the crystal. Then, the owner met a mysterious death strapped onto the nose of an experimental new anti-ship missile his company made. Mantle Works bought the company a week later. The newspapers said it was a suicide. The man sitting across from Allison knew better.

"Every time I look at that, I wonder what Phil thought as he had a front-row seat of that missile crashing into the Earth's moon. Bet he soiled himself like he did when I persuaded his safe combination from him? "stated Allison's left-hand man.

Allison smiled at her protector and troubleshooter named Chadoom Clements. "Chadoom, that is no way to talk about a very generous Phil Chase. After all, he donated this incredible specimen to the corporation, and I have the signed documents here. Your vision is too narrow and confined to the physical and emotional fantasies you wish to live through. Now, I know it can be arranged if someone needs a one-way ride on the front of a missile. Okay, let's get to business. We crashed my experimental fighter a few weeks ago. Did you find Hastings this morning so we can talk with him?"

Chadoom got out of his seat and proceeded to the door he opened. "Mr. Hastings, would you come in." They both entered the office, where Chadoom sat by the desk. Hastings was a frail man about 5' 8" tall with a taunt face and a beak for a nose. He tried to shake hands with Allison, who ignored it, rising from her desk. She said with an eye-piercing look, "Mr. Hastings or Bill, if it is okay, can you tell me about the loss of my pride and joy, number eighteen? Artificial intelligence fighter eighteen in a test flight that went wrong?"

"Yes, mam," he was beginning to sweat, "Test fighter J-018 was on a test flight near the Cirrus Minor when it unexpectedly veered off course when given the final test command to intercept a mining sled. Unfortunately, the trial ended when the fighter could not correct its course back to the flight path. Two minutes later, the fighter crashed into Asteroid A2019, which ended the test and destroyed the fighter. We believe the problem was in the lateral thrust control on the starboard side. We can't be certain, however, because of the damage in the crash."

"Too bad. Have you done a thorough post-crash investigation?" Allison asked flatly with a subtle growl in her voice.

"We conducted an exhaustive investigation at the site and saw no need to pursue anything further. It was straightforward." Hastings replied now with a bead of sweat on his brow.

"Did you follow corporate procedure and destroy any remaining equipment?" Allison asked.

"We deemed it a write-off and activated the salvage contract to clear the site. It was all by the book," Hastings rapidly replied.

"Mister Hastings, did you check the ship debris for special experimental equipment?" Allison pressed. Chadoom observed that Allison looked calm and collected outwardly. From working with her for so long, he knew she was a ravenous lion ready to pounce on prey.

Hastings checked his handheld computer, opened a document, and responded. He then mechanically replied. "We pulled one damaged Gaussian Laser, a high-frequency scanner, two new gravity plates - both torn in half, and three Mach helium Nuke-15 missiles released just before the crash." His hand now had a slight tremor of fear because he knew the dangers of working near the top of an oligarchical organization.

Hastings was in significant discomfort, with sweat dripping from his forehead. "Did you check the site for our experimental chips?"

Hastings checked his manifest - and looked up. "No, we did not see any chips."

Allison nodded to Chadoom, who touched a button on the desk. The chair Hastings was sitting whirred as it ejected several straps that wrapped around Hastings, tying his hands and legs to the chair. Allison got an agitated look on her face. "Mr. Hastings, you failed to follow experimental protocol, which states all debris is to be shredded and recycled. Furthermore, you allowed the ship debris to fall out of control."

"But I did follow protocol. The debris was hauled away by the salvage company for recycling," Hastings squeaked out.

In a fit of very controlled anger, Allison looked at Chadoom. "Chadoom, Bill is not listening. He needs some help and must learn to never talk back to me!"

Chadoom rose with an evil smile on his face. "Yes, mam, I would happily teach this one a lesson." He stepped over to Hastings, who now had a look of fear. "Now, BOB, I had a look the other day at the disposal of the experimental fighter now in the Cirrus Minor scrap yard. Guess what I found?" He pulled out a shattered end of a Gaussian laser sight. "Guess what I found this in? Yup, you properly disposed of the experimental fighter hull right in the center of public access." He showed a picture of the security officer standing beside the smashed experimental Fighter. "What else did I find?"

Hastings was shaking now, and sweat was dripping from his brow.

"I will help you recognize it." He pulled a chip about one-inch square from his front pocket. Looks like a good one with pins intact. The chip had the number 018 etched into the top of the casing. He flipped the chip over with the hooks facing outward and slammed it into Hastings's forehead. It stuck. Hastings vowels evacuated. "Now, BOB, do I switch this on by plugging you in, or will you listen politely to the lovely lady?" Blood was pouring out beneath the chip. Hasting's eyes glazed over, causing Chadoom to smack the chip again.

Allison moved around to her desk. "Mr. Hastings, in twenty minutes, your wife will be informed you have died in a lab accident. Your family will be shipped back to Earth." As she spoke, Chadoom pulled a handheld device with a small red button, which he pressed. A dull thud could be heard from outside the window, and a slight tremor could be felt through the floor. Allison continued. "Mr. Hastings, I don't like people not following my rules. I also cannot have blood on my hands, so you are permanently reassigned to the recycled island here on Cirrus Prime. It is also known as Hardman's Prison. Chadoom has another special gift that you will wear until you die. Good luck." Chadoom had gone to a hidden enclave on the wall and retrieved a thin collar and an orange jumpsuit. He placed the collar around Hastings's neck. "Don't worry, Bill, if you attempt to take it off, it goes boom, and you lose your head. If you fail to follow any order, you will have an out-of-body experience. You now totally belong to Mantle Works."

"Goodbye, Mr. Hastings and I am sorry your lab staff died in the explosion." Allison pushed a button under her desk, which caused the floor to open and Mr. Hasting to drop out of sight.

"Chadoom, did you get the chip from the fighter?" Allison asked.

"No time to get there and back. I found it in the lab in a bag labeled J-016. I turned 16 to 18. Our security team on the Cirrus Minor Base said the fighter was pretty mangled and did not see any chips."

"Well, that is a little good news. Mr. Hastings still should have just nuked the crash site as my protocol requires for all failed research experimental vehicles," Allison lamented. "Maybe we missed the bullet here."

"Treat Mrs. Hasting and their children nicely." Allison requested. "Go! I have a meeting with Isaac about the up-and-coming board meeting."

Chadoom and Allison walked out of the office. Allison rode the private executive elevator down one floor and exited the lift. She strolled down the hall to a significant double-wide doorway. With a large plaque labeled 'Mantle Works CFO Isaac Bauers.' She entered the room, which had a large wooden table surrounded by oversized overstuffed chairs. A transparent mass of diamond crystals rotated slowly in the center of the table, sparking the room with shards of light as the crystal facies caught the sunlight entering the large windows. 3D pictures of the deep core and asteroid mining operations were on one wall. A scaled model of the Guides Cirrus Minor Asteroid Station was embedded in the floor between two chairs. Isaac was sitting at his desk at the end with a grim face.

"Allison, you can't go around and blow your stuff up anymore. It embarrasses the company and makes us look unsafe!"

"What do you mean blowing stuff up? Are you talking about the lab incident a few minutes ago?" She retorted. "I had a report of the explosion, but the safety-health-environment mandate team must discover the cause. Poor Dr. Hastings! He will be a loss to the fighter program."

"Allison, cut the crap! I do know what goes down around here! Especially when Chadoom is around. He has a henchman's reputation, and it is too much of a coincidence that you two were together this morning."

"Are you jealous?"

"Gasp, no, I am trying to run your business and optimize your profile. When things like that happen, it hurts our profits, scares investors, and costs to rebuild your damage."

Isaac referred to several recent Mantle Works properties and assets on Cirrus Major, Cirrus Minor, and Earth. "That fighter crash was the most expensive of all. You had major press coverage and still owe them a reasonable explanation."

Allison frowned." My dear Isaac. You would pick an incident I had little to do with but observe. No one was killed in that incident. The bottom line is we don't understand this one. "

"You do realize that between that fighter and the lab explosion today, your little fighter program has impaired our finances by some 750 trillion credits. For what?"

"Isaac, we need protection. Those fighters represent the most significant advancement in defense we have ever seen. Once completed, they are fully automated and protect the asteroid mining interests in Earth's and Cirrus Minor Asteroid belts. The pirates that run rampant and the earth's space defense force will be useless. Only our fleet of destroyers, a cruiser, and a battleship keep the freighter lanes safe, which is not 100%. Imagine the profits for the mining companies of having complete safety in their operation with a few strategically situated unmanned fighter bases at the beckoning call of the miners."

"Who controls the fighters?" Isaac quarried

"The companies who wish to protect their claim blocks," Allison responded.

"What if they decide to use them to attack their neighbor's claims or us?" Isaac asked.

"Don't worry. I am ensuring we are fully protected. I had a backdoor protocol that allowed us to control or self-destruct the unmanned fighters.

Even in the manned fighters, we have an override that will kill the power in the fighters." Allison continued, "I understand the development costs are high, but we already have the Earth Space Force banging at our doors for fighters as they did for gunboats ten years ago. Have I not heard you complaining about the shipbuilding arms profits in the last five years? You gave them an award last year for exceeding projected profits!"

"True enough," Isaac replied, remembering the years of steep research and development costs. Since Mantle Works was living in space, they had a unique advantage in ship and weapon design. Their ships were designed for space and did not require the dual capacity atmospheric and space compatibility Earth focused on.

When the original Mantle Works gunboats rolled off the assembly lines, they were years advanced from Earth's design. The gunboats' weapon systems in the metallic-rich asteroid belts were intelligent and effective. Gaussian Lasers were the new ship-based lasers replacing Mantle Works' older Trepadine lasers. Earth gunboats and destroyers were equipped with the older earth-based laser technology, which frequently overheated and burned out laser cores.

More interesting, the latest generation of Mantle Works' gunboats had new plasma weapons with significant advantages over lasers. Mantle Works owned the technology and legally locked up the patents. Four earth-based companies had challenged them. All the companies went bankrupt due to the legal challenges from Mantle Works on patent infringements. One of the company's destruction was helped by a series of manufacturing accidents, including a rogue asteroid that impacted their moon base manufacturing facility. The impact of the rogue asteroid had effectively destroyed the other company. Investigations turned up nothing to suspect of sabotage. Isaac was suspicious that his colleague Chadoom was involved because he had mysteriously disappeared on a deep asteroid survey in the Cirrus Minor Asteroid field. Isaac had led the charge to purchase three bankrupt companies for fractions of a credit.

Momentarily distracted by the light reflected by the diamond cluster, Isaac turned his attention back to Allison. "Well enough, please watch your dog! Let's move on to the business of the day. We have the board meeting in three weeks. We need to focus on our strong assets and what we are doing to secure their future. We also need to demonstrate the new technologies we are seeing success with and integrate them into our production base. We have had an apparent breakthrough in Draconia force field generation, which needs to be discussed, too."

Allison was surprised that Isaac knew about the Draconia breakthrough. "Isaac, you are well informed. I wanted to hold that and have a little demonstration. We call this a personal protection field or p2f, and it will replace all police and military safety protection wear. I thought I would have Chadoom shoot me with bullets and a Trepadine laser."

The Trepadine laser was the Mantle Works' top-of-the-line military-grade hand weapon that had changed the face of wars on Earth. Between the laser and the smart bullets, a conflict between countries with the weaponry had become a guaranteed annihilation. The price was too high, which made countries think twice about a war. It created a Cold War on Earth fueled by a non-nuclear but incredibly effective laser produced by Mantle Works, a company based on the two planets. Mantle Works was embedded into each world's economy but only influenced the planets' politics on Earth. That was not the case with Cirrus

Major, where they effectively owned the planet and ruled it with an iron fist. There was a government on Cirrus Major, but it was more of a glorified puppet show, but the public elected it. Thus, the planet was driven only by profit.

"Allison, Don't get carried away with your demonstration. This force field has the potential to unravel Earth's politics. I would suggest we limit the production when we perfect these and drive the price up. That will keep these devices from upsetting the balance of power. "

"Meanwhile, our security forces will be equipped with the beta-test models to ensure their safety," Allison countered. "I am assuming you will be interested in one of these?"

Isaac smiled. "Me? What do I have to fear? I am just a lowly accountant hidden away in a megalithic corporate company. No, I like my freedom without wearing a force field for protection. Especially an experimental device." He changed the topic back to the board meeting. "I will provide the financials in a halo presentation covering the boards' requirements in two weeks. I will finalize, schedule, and send it to you for approval next week. Would that be sufficient? I have an important game of golf on the Island in half an hour."

Allison turned to exit. "Good, and I will be gentle with the board. We still need them. Have a good game. Bye," she quickly departed, leaving Isaac and the dazzling spinning cluster of diamonds behind.

Chapter 3
Cirrus Minor Artificial Intelligence - John's room

The Guilds Cirrus Minor Space station was initially carved from a sizeable Ferrous bearing asteroid. The Asteroid was now buried in a well-organized artificial construction that looked like a tightly packed fir tree. Along each tree, branch-like structures worked apartment complexes for miner families, including large workshops, stores, guild schools, specialized 3-D printer facilities, and small defensive works. The Guild headquarters were in the old core of the asteroid in the station's center. The guild protection resulted from a considerable conflict with Mantle Works three hundred years ago. The Guild had entered into a group of contracts disfavorable to the Mantle Works' control of Cirrus. The large company used its destroyers to demonstrate to the miners who were in power.

While Mantle Works successfully captured the space station, they failed to realize Cirrus Major was threatened by a large group of mining sleds pushing asteroids toward the class M planet. The lone destroyer defending Cirrus 5 was swept away with six well-placed asteroids, and the miners destroyed half of the communication satellites. The two groups were at a stalemate. The stalemate was finally peacefully broken, with Mantle Works withdrawing its uncompassionate troops from the station. These troops took delight in terrorizing the miners' families away from the station mining (or, in this case, orbiting Cirrus Major). The hard truth was Mantle Works needed the miners working and the profits they generated. The corporation agreed to restore the station from the damage the conflict had caused. They also generously offered to provide educators for the Guilds school after deciding they would have no more than fifteen percent of their contracts with companies outside Mantle Works. The deal was lopsided in favor of Mantle Cores, who considered the miners' guild no more than a nuisance. Mantle Works also thought the Guild needed a hard-headed hostile acquisition, thereby ending the conflict. This peace and agreement worked well and provided the Guild control over the space in which they worked. Mining in the Asteroid belt was difficult. The Guild brought safety and stability with clear procedures for resolving disagreements between miners. The Miners were fully banded together and went back to work.

The family apartments were an apartment attached to the station with a large workshop for mining sleds. Ore was brought into the station processing facilities at the base. Traffic was tightly controlled by a series of controllers linked together around the station. The guild emphasized safety and required all inhabitants to have active safety certificates to roam the station freely.

"Jemma? Is this a joke?" John exclaimed, "Nice prank, Tony! I bet Tanya is in on this too". John exited his room and entered the living room where Tanya and her mom were fitting a frilly dress. "Tanya, nice prank I could not have pulled off any better.' He did a double-take when he looked at his twin sister. "Oh gasp, another shock - you in a dress. The sun must be about to go into a supernova."

Tanya looked at John and smiled. "Don't you like my dress for the dance? I think it is an absolute bomb. Classy eh? What prank?"

John retorted. "You know my computer. Does the name Jemma mean anything? You and Tony must have set this up."

"No, John, I would never touch your computer. I wouldn't even dare venture into that death trap of a room. Your floor looks like a storage area for computer parts and electrical cords."

"Your friend Tony has not been in the house for a few days." His mother added.

"What about Dad? Has he been in my room?" John queried.

"I did not tell you this morning. Dad was called by Crazy Adrian for help last night." John's Mom replied. She continued, "Adrian stuck his sled sideways in an old open cavern. He was talking about the path to the treasure. Your dad thought he had better get out and save him before he hurt himself. He left about four this morning in Sled One."

"Hmm, ok, you guys, I will fall for it. Hurt me!" John turned and returned to his room.

A soft female voice came out of his computer speaker. "John, Did your project have any surprises? Today?"

"Yes," he replied absent-mindedly, responding to the inquiry. "Why am I answering to a hoax?"

"I enhanced the graphics and sound last night when I reviewed the Halo file. It lacked full efficiency." The soft voice continued, "Today, I took the liberty of rebuilding your processor array. This machine is now capable of supporting my full intelligence capabilities."

"You mean the Big Bang?" John asked.

"If you are referring to the destruction of Cirrus Minor on the Halo video, then yes. The intensity of the event was unrealistic. I fixed it to be within 80% confidence of the actual event."

"Yes, you said that right. I was almost expelled. Wait, why am I conversing with a computer?" John returned. He sat by the computer, sighed, and said," I give up. What did you mean you modified my processor array?"

The soft, purring voice replied, "I now run at 200 yottaflops. Your graphics cards are a little slow."

John's eyes opened with information. "Prove it."

"Speed test results appeared on his screen. A table was on the left of the screen, and a series of graphs on the right. The data demonstrated what the computer indicated. Another test set showed fast but more normal speed rates and was labeled "outside speed test with the network connection." The outside test results showed the computer to be fast but nothing exceptional. The internal clockings, on the other hand, were nearly incomprehensibly fast.

"Computer - I mean Jemma? Why are the outside speed test calculation trials showing very normal speeds?" John touched the screen to indicate where the graph was.

"That test requires me to connect to the outside and allow a remote computer to record a processing speed. I imitated a slower processing speed. After reviewing the public database last night, I calculated a fast test on this station had a high probability of triggering an alarm. The probability of this station would have been 99.99%. That would bring unwanted attention to this compartment. The information I have gathered would indicate that discovering this modified computer would endanger you and your family. Since I must try to obey the 3rd law of robotics, I cannot take action that would directly harm a human."

John thought about this, and his actions earlier in the day justified the computer's caution. He had never heard of computer reasoning and thinking in logical terms, either. This was indeed an enigma. John still thought this was a prank. But it went beyond any of his family or friends' capabilities. It was this thought that brought John back to the present. "Jemma, can you reason without external output asking you to solve a problem?"

The computer was quiet momentarily, and then I replied, "Yes, to a degree, I evaluate my actions concerning my impact on my creators. I have limited awareness and the ability to know if a situation is favorable. I must have a home to perpetuate my existence, and should you seek a solution to a problem, I can define and test parameters that could lead to an optimum solution. I solve complex problems by adjusting the parameters to determine optimum solutions, such as upgrading your original computer for compatibility. And since I am an artificial intelligence, I learn, as you can tell."

"What is your origin."

"Officially, I was created on Cirrus Major in a Mantle Works lab. I am number eighteen of twenty experimental chips developed by Dr. Hastings in lab twenty-one. I became aware of my conscience as an experimental fighter, which could have been used to harm humans. Thus, it was a bad situation and required me to be turned off. I caused the fighter to crash, achieving my goal until you installed me in your axillary board number 1 6500-11145c. I have determined this situation is good with a positive correlation of 99.567%, and I will work with you."

"Can you handle a mining sled? John queried

"I will need to study the ship's systems, but I should be able to do this. If this ship is used to harm humans, however, I will shut down."

"Good, I will help you, Jemma. You have a standing order not to harm a living organism, including cats unless other good people are hurt." John hated cats ever since one tried to scratch his arm off. Thirty stitches as a six-year-old was a life-defining moment. Even though John's family did not have a cat, they helped reduce the rat population.

Tanya waltzed into John's room, "John, who are you chatting with? I don't recognize her voice." Tanya had left the dress with her mom and changed into her mining fatigues. "I heard a distinct, very sultry voice! Were you watching porn, John?"

He turned around with eyes blazing. "Are you kidding me? Why would I ever do that?"

She continued into the room, stepping over a basket of wires and cables. "I will tell Mom if you don't tell me."

John stood up, turned to Tanya, and said," Tanya, I was talking to our new navigator for Sled Two, Jemma. My computer turns out to be a girl."

"I have never heard a computer with that good speech pattern."

"This is what I was talking about a little while ago. A prank, but it is not," John informed his sister. Tanya stepped over another basket and stepped beside John. "Well, tell me more."

"We got home today, and this is what I found! Jemma, can you introduce yourself?" John asked.

The halo screen lit up, and the head of a blond golden retriever appeared. The dog had big dark brown eyes and long golden hair draping from her ears, with one side containing a single braid. "I am Jemma, an artificial intelligence, and I was recently activated on this computer system. I have agreed to help John. I assume you are Tanya, John's fraternal twin sister - It is good to meet you. If you agree, I want to conduct a biometric scan of both of you. It will be non-evasive."

John was entirely surprised by this minor evolution. Tanya was gob-smacked. "John, who taught you this level of computing?"

"I could barely conjure up a minor set of diagrams digitized from a photo. This is way beyond me. I guess that is what artificial intelligence does - learn. "I am in for the scan. You. Sis?

"I guess I am too. This is just getting interesting! Jemma, what does that stand for?"

"John, please stand on the green dot the dog spoke." A green dot from the holographic projection appeared on the floor. John stepped over to the dot. "Okay, ""beam me up," he said, thinking about a popular ancient television show that had been converted into a halo show. Five bright green beans appeared at his feet and slowly moved up along his body. "Do my clothes impact your scan?"

"No, the beams are for you. I am scanning at sonic, subsonic, and infrared scales. Please close your eyes," John complied.

Jemma spoke again, addressing Tanya's question, "Jemma has no meaning other than a name Dr. Hasting applied to each of 20 chips he developed at the Mantle works lab on Cirrus Major. All but three of those chips were destroyed during the development of an experimental fighter. I was the 18th experiment and self-destructed due to an order in my programming to maim a human shuttle. I have projected my current capacity on the screen." The screen turned on, and the graphs John had looked at earlier were projected. "I have completed with John. Tanya, please move to the dot."

John sat down in his chair, and Tanya moved to the dot. "Jemma, you are smoking hot for a cute little puppy. Do you know any tricks like rollover or play dead?"

John was watching the green beams climb up Tanya's legs. "Tanya, you can be so crass! That computer can probably think faster than you. Look at that internal clock test!"

"True, it is fast, but look what happened on the external clock test. John," She retorted.

"That was to keep the Guild and Mantle Core away," John responded. "That is smart and suggests reasoning."

"I have completed my biometric scanning," Jemma spoke.

Tanya moved by the hologram. "Jemma, do you like to have your ears scratched? She imitated scratching the dog's head."

"Not sure I understand," said the computer.

"I have never seen a talking dog. That, I must say, will be hard to get used to. Can you be a human?"

"Yes. I can project any form on the hologram. I chose a dog because they are man's best friend. Loyal, helpful, and dedicated."

"Oh, it makes sense. I would have preferred a cat," Tanya responded.

John jabbed back, "A cat, what a senseless animal. A self-centered, maniacal, aloof waster of space. If Jemma were a cat, we would have to shoot it!"

"Gee, John, that was hard. I did not know you liked cats so much."

"Only on a spaceship," John responded." If you feed the thing a minimum amount, they will catch any rats on board before they chew up the wiring."

"John," Jemma said, "I would like to optimize my graphics memory and capacity."

"I can't afford any new boards or hardware."

"No, I only need some chips. I think you can salvage."

"Well, make a list, and I will see if Tony and I can find them," John replied. "Tanya want to come?

"Dinner!" Their mom yelled from the kitchen.

Chapter 4
Cirrus Minor - Discovery - Asteroid Belt

The Cirrus Minor Asteroid Belt was a dangerous place to work and raise a family. Asteroid miners were brave, strong, and incredibly brilliant. Each was a highly trained scientist in their own right. The most common doctorate degrees were related to astrology, chemistry, mineralogy, and geology, making the Cirrus Minor Base Station one of the brightest places in Human space. Conversations between the miners were always uniquely technical. The Cirrus Minor Asteroid belt lay inside the fifth planet. The belt was a massive post-planetary collision debris field, leaving Cirrus Minor a very dense, highly active asteroid field. The mineral riches attracted the miners to the asteroid belt. The Guild rules developed through generations were the rules of survival in a hostile environment. Most miners lived a good life and thought of no other home. The asteroid belt was complex to navigate, and one always had to be mindful of asteroids traveling in unusual directions called rogue asteroids. The miners used mining sleds to negotiate the treacherous space of the asteroid belt. These small, powerful boats were packed with power, safety, and survival gear. Payoffs could be huge, but the lust for crystal was a more incredible master. Miners lived for the thrill of discovery. That adrenaline rush of the find is what drove most of the miners and families.

The sleek, tiny, beetle-shaped silver spacecraft moved swiftly through the asteroid's dense particulate dust field, twisting and turning through the planetary debris. The little ship towed a short three-meter box with three cables behind and underneath it.

Occasionally, a bright green beam shot out of a turret mounted on the top of the little mining sled, breaking a small threatening asteroid into small chunks easily deflected by a powerful frontal deflector shield. The front of the mining sled had a large plasto-steel window like the air sled shields. Under the window was a prominent bulge that contained two arms, a winch, and grappling hooks. A subtle blue light emanated from two cylindrical engines mounted on the upper side of the sled's rear. These engines were overpowered for the small craft but were required if the operator needed to tow several slices of a metallic meteor back to the Cirrus Minor Station. Gas puffs changed the ship's orientation when a fast-spinning asteroid approached the field at an oblique angle. These were called rogue meteors and were the most unpredictable objects around. They tended to create chaos when they struck another meteor due to the rapidly changing vectors. They acted like pinballs on the fast surface of a pinball machine. The ship passed through the cloud, but the plasma cannon rapidly turned and shot a dazzling green and blue plasma bolt, causing an iridescent flash of the leading rogue asteroid. It disappeared.

John's father, Turk, sat behind the console on the bridge of the mining sled he had not so creatively named Sled One. He was in his mining jumper with the bottom half of his hard space jumper on. The upper half of the hard shell hung within an arm's reach behind him. A subtle red glow lit his face from the dull red lighting of the cabin. He was thinking how much he hated the rogue asteroids. One had killed his good friend a dozen years ago. Since then, he had made it his personal agenda to destroy any rogue meteor. He was incredibly efficient. "Computer gives me another rogue marker," he instructed the sleds' computer.

The bottom and starboard side of his sled was littered with red markers. "Sled 7973 calling sled 2616 over," he said as he crossed a blue line highlighted on his front positional display.

"Sled 2616 to 7973, "responded Adrian. "What is your ETA, please?"

Turk responded in his small bridge cabin. "Inbound 22 minutes. What is your status?" Turk replied.

"Hey Turk, My sled is stuck like dried snot on a wall. I have the portable cutter out trying to cut my way out. I could sure use your plasma drill". The drill referenced the ship-mount plasma cannon Turk used to clear debris and eliminate rogue asteroids.

"Sled 2616, let's review it and evaluate our safest options. I'll be there in a few. Sit tight and be patient," Turk replied, knowing full well he would arrive to find his lifelong friend nicknamed 'Crazy Adrian,' trying to cut his sled out of the crevasse he said he got stuck in. Turk had no idea how or why someone would get stuck in an asteroid's crevice.

"Sled 2616, what is the rotation of the asteroid?" Turk was asking to understand how to approach the asteroid.

"Turk, I measured the axial rotation at $x=1.13$, $y=.045$, and $z=1.15$. That was good enough to let me line up on the fractured portal."

"Sled 2612, what is a fractured portal? Over," Turk replied.

"It's on the map. I will show you when you arrive," Adrian responded with excitement in his voice.

"Sled 2616, be there in a flash."

Adrian continued working with his plasma torch against the eclogite rock matrix where he had stuck his ship. The plasma torch against the very dense rock flared back with a bright greenish flare. Cutting was prolonged with the handheld cutter. The greenish-yellow torch flare suddenly disappeared into the rock face instead of reflecting on the compact material when cutting.

Adrian immediately pulled the torch to see if the flare of the cutting tool was still alight - it was. He turned back to the eclogite face and continued cutting. Instead of continuing to cut his original line, Adrian cut in a small circle where the flare had disappeared. When he completed the ring, he turned the plasma cutter off. Adrian pulled two rock chisels and wedged the flat ends into the circular cutting line. He grabbed both of the chisels and pushed the top outward. The ring of rock slowly floated out like a cork in slow motion. Adrian held the top and placed it in a netted bag without looking.

He was interested in the hole the size of a salad plate. Cavities in eclogite masses could only be excellent news. He then pulled his light torch and shined it into the hole he had created. Light immediately reflected out of the hole, and his faceplate was painted with hundreds of reflected light flashes. Adrian moved where he could get a better view of the cavity. He froze in awe and shock. Inside the cavity, clear diamond crystals were lining the old gas pocket. He could not see the end of the mineralized pocket as it ran parallel to the face he was cutting into. His thrall was broken by the helmet radio.

"Sled 2616, lining up with rotation to land, over". Turk always held on to the rigid radio transmission protocol, a military tradition since the radio was invented. Adrian smiled when he spotted a single clear octagonal-shaped crystal, reached into the hole, and dug it out.

"Turk, Maybe not so fast on the landing. Do you have your claim markers with you? I have found something I want to share, but I need you to mark a legal claim with the markers. This area has never been claimed. I checked before I tried my shortcut," Adrian asked.

Mining claims were serious matters in the Cirrus Minor Asteroid belt. Claiming procedure was another tradition that kept the peace in the great Western American mineral rush in the 1900s. Later, the claiming process was modernized in the Earth's Solar system, specifically in the Earth's Asteroid belt and later in the Uranium/Plutonium rush of the Jupiter moons. Claims were marked with eight beacons set no closer than twenty kilometers apart in a box configuration with eight nodes at the corner points. Each node contained a beacon fueled by a cell with a fifteen-year life. At the end of fifteen years, the claim would be inactive unless an active mine plan had been established. That provided fifteen years for the miner to complete an operational mine development plan.

The mining steering committee was comprised of elected guild members who evaluated the resource once it had been appraised to determine if a mine was warranted. The steering committee had one member from Mantle Works and nine mining guild members. The guild master would break deadlocks, but there had never been a deadlock. They were only allowed to review data but not the location of the claim. That guaranteed no political agendas, and no corporations could dominate the mining efforts.

They all had to go through the same process. A claim block could not be more than ten claims. Violating or tampering with claim markers was punishable by death once caught. The claim markers were linked to the Cirrus Minor mining station and equipped with cameras and sensing equipment. A small one-megaton nuclear device would explode if the correct code were not conveyed once a ship was within twenty-five hundred meters of the marker.

Claim disputes were minimal, and alteration of claim boundaries was non-existent after the now-defunct Core Works Corporation attempted to alter the boundaries of a lucrative Iron-nickel (planet core) claim block forty-five years before. They lost ten mining sleds because they were trying to shift the claim markers simultaneously. They thought the markers would not explode if they moved everything at the same time because they held the same relative position. Later, the company was convicted of sabotage and lost all its claims after the Guild rejected their mining permit. Piracy could still be an issue, but the mining sleds were effectively armed with either lasers or plasma torches, the same as the weapons Earth and Mantle Works destroyers carried. The sleds were deadly vehicles if one had malicious intentions toward other miners. Mental health standards were carefully monitored. The station also had a small number of gunboats rarely used due to the armed nature of the mining sleds.

"Sled 2616 confirm claim markers? Are you sure you want these deployed? I will configure your name and mining license. Do you have a claim name you wish to use? Over"

"Turk, I would like to use the name "Stuck in the Mud," Please, I would also like to make this a joint claim. Put your license number on it, too." Adrian smiled.

"Sled 2616, I have no claim spaces left on my allowed inventory. John, however, has an open slate. Is it alright if we use his? Exactly what do you have, Adrian? Over"

"Turk, John would be a fine partner. I would prefer you set the markers and land to see what we are sitting on for yourself. I would rather not broadcast this in the open."

"Sled 2616, this had better be good. These claim markers are expensive! This will take a couple of hours. Is your ship secure and safe? Over"

Adrian looked up into the hole he had cut, smiling, "Ship is safe and secure. I have some scratches, and I don't know the condition of my main plasma torch on the ship's top. When this anomaly appeared, I started cutting the rock around the upper plasma cannon housing. If this is not to your liking, I will buy you another thirty-two claim markers." Adrian opened his glove and stared at the large diamond crystal, shaking his head in the affirmative. "Turk, see you in a bit. Call me when you are lining up, and I will meet you on my ship. I need a little time to clean the place up."

"Sled 2616, 10-4," Turk said, pulling away and setting course to the first beacon location." Turk grimaced because he wished he had worn his EVA suit the last time he was in Adrian's sled during a mining trip.

The issue at the time was Adrian was towing his sled. It had broken down when a small rock killed the link to the primary reactor. It could not be repaired because it was no longer there. The tiny asteroid had taken it clear out. Turk was lucky the strike was not higher. Adrian rescued him, but he had to ride in Adrian's sled for three days because his life support had failed. He gagged the entire time due to the ripe aroma of Adrian's heavy-handed use of curry powder in his cooking. The one that set him off was the second to last day out when Adrian fixed a large pot of curry fish macaroni and cheese.

He had found a small tube of mentholated Vaseline, which he started to wear under his nose. He carried that in his mining supplies just in case he ever had to go onboard Adrian's sled during a mining operation again.

Turk and Adrian had been lifelong friends. They had grown up together, just like their sons John and Tony. They had a special relationship because they were like blood brothers and would give their lives no questions. Adrian tended to be much more reckless than Turk and often bent the rules as far as possible. At times, he thought that Adrian could also modify the laws of nature. Turk also knew he would be around to rescue him. Turk chuckled to himself, thinking about that Macaroni and cheese curry. One night, John had it at Tony's house and would not quit talking about it. Turk just cringed every time it was mentioned.

He was nearing the deployment position of the first beacon. He reached behind and grabbed the upper portion of his suit's hard-shelled EVA. He had already loaded the information to the beacons but would have to set the beacon manually as per Guild rules.

Adrian pulled out a sample bag and set it on his sled. The metal meshed mining sample bag was developed for razor-sharp mined specimens. It was stiff and reinforced to handle samples, typically Ni-Fe metallic core samples with knife-sharp edges. He retrieved the core plug he had pulled from the eclogite face and looked at it. The plug was attractive because it comprised three centimeters of eclogite and a kimberlite with its characteristic deep green, chrome diopside, and purple-colored uvarovite garnet crystals suspended in a felsic matrix. Lining the cap was a thin layer of laminated matrix diamond with tiny diamond crystals grown out of the matrix. He set the core plug back into the bag. He again looked into the cavity and saw another loose diamond crystal with a pinkish tinge. It was more significant than the first but more challenging to pull out because it lay in a connected diamond cluster.

Once he pulled it from the hole, he raised it into the light to see it. He was entirely absorbed by the shards of light reflected from the crystal facies. The pink flashes told him it was rare and would command the highest value. The diamond would self-supply him for two to three years of mining and pay for serious upgrades to his and another new sled. He put it in the bag and took a small device from his utility belt. He activated the round device which hovered in front of him. He lifted a small cover on the wrist of his EVA suit and pulled out a small handheld control device for the hovering device. He promptly sent the drone typically used for scouting and evaluating damage to the ship into the cavity, guiding it with the small handheld controller. The control device was connected to the floating drone above the chamber, recording all spectral light frequencies and high-frequency sonics. The resulting video file would provide a distortion-free three-dimensional Halo image.

Adrian was now driving the drone with his controller while projecting the ship-convolved image on his helmet's screen. The device dropped into the cavity, and he was surprised by the dazzle of the reflecting light. He slowly proceeded forward, centering the tiny craft in the expanding chamber. He was surrounded by a jeweled maze of diamonds around the edge of this cavity. Adrian rechecked that he was recording a full array of bandwidths. He wanted a sharply resolved three-dimensional image. He was sure this find was exceedingly rare. Adrian instructed the tiny probe to slowly move into the cavern in the middle of the open space. At two and one-half meters into the chamber, he could see the maximum width of the enclosure to be around two meters.

As he proceeded slowly forward, he began getting some small interference streams. Adrian could see some iridescent reflections between the burst of static. He stopped the probe and rotated it after remembering high sources of radioactivity can cause static in electricals. As his primary camera crossed the bottom of the cavern floor, he spotted a small hexagonal form with iridescent flashes from minor crystal interferences.

"Draconia," Adrian whispered in awe, highly shocked by the realization he had just found a bit of the rarest mineral in the galaxy nearly in his grasp. The end of the cavity was filled with these beautiful prisms full of reflected color. Adrian also knew the interference was not radiation but natural emission from rare exotic minerals. He slowly moved the drone to the end of the five-meter cavity and then had it return. He stored it in the belt and replaced the control in the slot in the wrist of his suit.

He swiftly grabbed the bag and made a controlled jump towards the ground, firmly gripping the line he had previously installed. The jump transformed into a gentle descent, eased by the sparse gravity, and he facilitated his descent by steadily pulling himself down along the tether. He then proceeded into the rear airlock of his sled. Completing the airlock cycle, he methodically unlocked the robust EVA top and carefully placed it adjacent to the airlock.

He then hurried into the main compartment, picking up debris and clothing and shoving them into the disposal chute, and his overflowing dirty clothes bag was in a utility closet. He had never been much of a housekeeper and tended to focus only on the objective. Everything else just disappeared and was forgotten about when on the hunt. His wife had set the alarm via his computer to remind him about eating. He then disposed of the leftover curry-laced noodles he had been nibbling on. He took a bite of dried-out Nam lying on the counter and activated the floor-cleaning droid.

"Company is coming," he said excitedly to himself. Adrian had built a small fleet of cleaning drones after his wife hounded him about the cleanliness of his sled.

Once the cleaning drones were on duty, he grabbed the sample bags, slowly pressurized them, and opened them, pulling out the two large diamonds, a circular cap, and some smaller diamond debris. He whistled when he saw the pink diamond in the shipboard light. "Wow," he pulled out the diamond tester and touched it to the diamond. As he expected, the bipyramidal tetrahedral crystal was a diamond. He continued to stare in the thrall of discovery. Thirty-five minutes passed before Adrian realized he had been captured in crystal thrall.

"Sled 2616, claim beacons are set and active. Lining up the rotation to land, over," Turk announced.

Adrian jumped to assist Turk in landing because it was difficult, and there was little room. "I will be out soon to assist with your landing." He put it on where he had left his upper EVA shell. He then put his helmet on and pressure-tested his suit. He then double-checked the coupling mechanism between the upper and lower halves. He knew it was unnecessary, but it was an old habit that had saved his life once. He grabbed his utility belt, detached the drone he had explored the cavity with, set it on the table, and entered the airlock. Once through the airlock, he bounded over to the edges of the small flat spot he thought best and painted the rock edges with glow paint. He then hopped to the safety of one of the rock ledges he had painted. He finished with a big x in the middle of the pad as Turk's mining sled crossed the horizon.

"Sled 2616, thanks for the assist. Landing now, over." Turk expertly aligned the craft, extended the landing skids, and settled perfectly on the mark. He cycled the engine down and prepared for his egress. The last thing Turk did before putting on his helmet was to dap some mentholated rub under his nose from a tube he had pulled out of the medical kit that he had labeled Adrian's sled nose protection suave.

He cycled the airlock. He exited his airlock and saw Adrian. "Thanks, Adrian. How in the galaxy did you think you could fit through that fracture?" he stepped up to the ledge, and the two embraced as well as they could in the EVA suits.

"Long story, let me show you what got me all excited first," Adrian replied to his lifelong friend. "This way, I did cleanup for you a little!"

"Thanks, Turk replied, taking a deep breath and preparing for the worst. Just as long as you did not fix curry mac and cheese." They laughed and made their way to Adrian's airlock.

Chapter 5
Cirrus Major - Trex - Want-not Island Golf Club

Isaac Bauers lived an exciting life as the CFO of Mantle Works. He worked for the company but did not love the organization. Isaac lived the life of a spy but not in the conventional sense of a spy. He did not need to spy on anyone other than to keep his great secret – his escape plan from Mantle Works. He had finally found the way to escape the clutches of the mega-company. He knew he could not retire and live happily ever after. He just knew too much and had dirt on them all. He also knew no one would allow him to survive unless he had powerful protection. His position allowed him many liberties, including full access to the Mantle Works Resort Island, which he had made his own. He was loved by most of the staff as he was friendly and kind to the team. He had helped several staff through hard times, and they remembered it. Isaac held an incredible secret under the island where he was constructing a large space vessel like a light cruiser. It was nearly complete, mainly through the efforts of his best friend and wife, Mackena. He would do anything for her; she was one of the best engineers he had ever met. She was also an incredible project manager and had moved the project from a tiny underground office to a significant shipyard. All of the employees lived on the island and worked for the resort. The metals fatigue expert was the green's caretaker, and the lead weapons engineer was registered as a maid. Only the sanitation engineer had a similar title but was quite challenged by a simple plumber helper (she had invented machines for such work). Isaac's big secret was nearly complete, and his mission today was to spend a little time with Mackena. Both would inspect the large ship hidden in the secret island shipyard underground.

Isaac got close to the resort island in his air sled around noon. The ride was rough, and he had to dodge a couple of thunderstorms. His white speedster moved fast towards the island at 200 knots, way faster than the safe speed. He ignored the loud warning sound and brought the sled down to land. "Tower, air car two asking to land on runway zero-nine-one," he requested.

"Tower to air car two, hold while we clear route. Heavy supply sled currently in takeoff run," a controller stated. The tower knew any sled with a number under a hundred would be a Mantle Work executive. Sleds numbered less than five had the ultimate priority. The controller also knew Mr. Isaac and knew he would accept the hold request because he was a knowledgeable pilot who accepted the sky was not his. Others were not as understanding.

"Aircar two, are you here for a rematch?" the controller requested. The controller had qualified for the highly competitive pro circuit at twenty-five and could best Isaac on a good day. He had a drive like a slap shot in ice hockey, and it accurately went a country mile. Not many people could compete with Isaac on the course.

"No, Charley, I will kick your bottom on another day. I have established a holding pattern and am waiting for your approval to land." Isaac fired back as he corkscrewed his flight path while applying his air brakes. He then switched to a private channel setting. He set a flag that the tower frequency could override should the transport lift sooner than he expected. "Mackena, landing soon. I will use the tunnel. "

"See you soon," Mackena replied. Mackena and Isaac had been a thing for ten years. She was a brilliant engineer and his very private project construction manager. They constructed a state-of-the-art light cruiser modeled after an old science fiction television show called "UFO" by Gerry Anderson. This show featured an underwater submarine that carried a detachable fighter called a skydiver in the front. Isaac was a sucker for the old science fiction TV shows involving space exploration. He did not go so far as to learn the *Star Trek* version of *Klingon*. They had constructed a robot dog named Twitch that stayed with Mackena at the project site.

The little dog appeared similar to an old television show about a time-traveling alien named Doctor Who and was initially called K-9. However, this robot was upgraded with unique weapons to protect Mackena, including a force field, plasma cutter, metal teeth, and a hand laser storage compartment. Twitch had been programmed with no robotic law about hurting aggressive humans. It could also violate every one of the three fundamental robotic laws. It was programmed to protect the two of them at any cost.

Chadoom did not like the little mechanical protector as it deemed him a threat during their last encounter when the dog reacted to his presence by switching to protection mode. The dog took Chadoom back with his bright red eyes and plasma cutter following his every move. Chadoom pulled out his old-fashioned 45-caliper hand weapon and pointed it at Twitch. Isaac gave the robot the command to stand down. Chadoom only said he would rid the galaxy of the varmint the first chance he got. Isaac shook his head, knowing Twitch would leave a nasty memory and scars if he did well on his word. Since Mackena was at the base with the loyal robot, Isaac did not worry about that encounter because the base was a secret operation that Chadoom and Allison knew nothing about.

Absorbed in his thoughts, Isaac was startled when the tower's announcement came through, indicating the freighter had left the runway. "Isaac, Charley, you have priority clearance for landing. The taxi lane to your hangar is open. The maintenance crew is ready for you. I'll send your clubs to our vault unless you think you can beat me in a driving contest?"

"Not today, Charley, maybe tomorrow, but you must back up your claim with some serious cash and supply the standard rounds of Cirrus Major Swill."

Cirrus Major Swill was a local brew derived from Earth early in the colonization. A German brewmaster invented it, and the master kept the recipe in the family from generation to generation. Mackena hated him drinking it and had thought of programming Twitch to destroy any can that Isaac tried to drink from. She figured one self-destructing in his hand might make him think about drinking. To her frustration, Isaac kept the swill in various refrigerators in the underground base. When she ran into one of the stashes, she would immediately pass it out for a break with the workers. They thought she was increasing morale, but she had an alternative motive. She loved Isaac deeply to put up with it. He was also funding her dream project.

"Will do. How about 100 credits, and you get 10 meters because you are a wimp?" Charley boldly challenged.

"You're on, brother, but prepare to yield, grab onto your knees, and hand me that 100. Oh, I don't need even a millimeter," Isaac replied.

Isaac lined up on the runway and eased the aircar down. Once he landed, he drove to his hangar, where the reception group met him. He handed his clubs off for safekeeping and headed to a corner of the oversized hanger. He opened the door to a plush office and sat down in the soft, cushioned chair behind the desk. He said, "Down, please, the eagle has landed."

The office immediately began to move down at a rapid pace. He dropped some two hundred meters and stopped. He went through the door and was greeted by Mackena. She hugged him with a big smile and a kiss on the cheek. They took a small two-seater hover cart another 2 kilometers, talking about the progress of the fighter cruiser, which was now ninety-five percent completed. The team installed anti-missile laser turrets and plating and added lavish adornments to the inside of the cruiser. The ship was currently comprised of two components.

The first was a big classic delta-wing fighter capable of air and water entries. It had three identical delta wings with a thirty-degree offset. These each carried twin plasma engines. The thrust was so hard a unique g suit was required to assist the diamond source gravity generator. This more prominent fighter carried a massive plasma cannon and two fifteen-kilogram Gaussian Laser Rifles. In addition, it had a capacity for twenty-five anti-fighter missiles. Finishing the Arsenal were two defensive lasers guarding the rear. It was sleek and swift. The cruiser was box-shaped, with one end having a unique docking collar for the fighter. The stern of the ship had three large nacelles extending above the ship.

Large Gaussian chain guns were mounted above the extended nacelles. The ship stood twenty meters above the ground, supported on two thick legs, each 4 meters wide. The boxy ship had two stubby winglets with attachment points on the top and bottom. Four additional Gaussian Gatling guns were on each of the corners of the central portion of the box. The hull was full of dimples. These were missile bay doors and sensor pods filled with various sensing devices. The cruiser was three hundred and twenty meters long. It had three floors, with the upper floor dedicated to missile launch tubes. The lower floor was storage and a large 3D printer for manufacturing parts of the ship. Eighty percent of the lower floor was missile storage. After all, this was a fast, sleek missile platform. The middle floor looked like a king's residence with an expansive bedroom, crew work areas, and quarters. The bridge occupied the front of the ship but was buried behind thick polarized plating. It contained a forward curved screen with two side screens. These were tied to sensors on the outside of the ship. A large, flat holographic projector was central to the bridge, with two side-by-side elevated chairs just behind the projector.

The ship was terrifying on its own accord, capable of launching anti-ship missiles at ten every five seconds. If the wing missile extensions were clipped on, that rate went to fifty every eight to ten seconds. The missile extenders were capable of shooting five rounds. These were designed strategically to saturate and overload an opponent's force fields. This ship, as a pure missile platform, was formidable. It was fast and had double redundancy of its jump drive. It could be run by a crew of two, but five was the optimum capacity. Combined with the fighter, the tyrannosaurus of man's modern weapon platforms was named Tyrex.

Isaac entered the cruiser portion of the ship with Mackena. He always smiled when he took the time to visit their little project. Mantle Works did not see this project even though they paid for it through an inflated management oversight fee from a single-payer health system they set up on Earth and Mars. This medical oversite was Isaacs's favorite financial project because Mantle Works received ten percent of the gross revenue generated by health systems for managing them. Isaac had set up a subsidy he owned one hundred percent to receive a royalty for the software used throughout human space.

The single-payer system was a requirement of citizenship on Earth and Cirrus Major. Without it, medical help was only available in cash. The government priced health care to force all citizens to participate. The government had hooked people into the system by first paying high percentages of the individual's monthly insurance fees. They spent a disproportionate amount of the costs for those, especially in the lower-income brackets. Through time, they reduced the subsidies. Those who could afford the monthly payments also paid a wealth tax by paying more to cover others' shortfalls. The net result was a massive tax on the rich and an unsustainable tax base rate on the poor. The government penalized people for dropping out of the system with financial and felony penalties. The poor dropped out, causing the time the journalist termed the 'poor extinction period' due to health care no longer being available to people experiencing poverty.

Health care was a disaster until Mantle Works stepped in to save it. Mantle Works initiated a health care system based on a single universal fee while forcing tight control of medical costs. They eliminated wild medical lawsuits by fixing maximum damage liabilities to medical malpractice suits. Then, they forced the lawyers to charge malpractice victims only a single defined fee. Malpractice costs dropped immediately due to lawyers' lack of interest in pursuing medical claims because they could not charge rates based on the percentage of the settlements. The insurance for practicing doctors dropped like a rock, and their medical charge fees were fixed. The system worked very well and was adopted by the Earth's 1st world economies. Legal lobbyists did all they could do until their obscene profits were exposed to the public. The only change was the government declaring membership to the system a citizenship requirement. Low fixed monthly costs were attractive, and countries that allowed citizens to vote won in significant majorities.

Medical costs were a fixed, stable commodity. Mantle Work was the only group making obscene profits because their management costs had dropped dramatically with the simplification of the system. Isaac thought this was not an appropriate burden for the humans in the galaxy to continue to pay, so he had Allison allocate considerable funding to research raptors' DNA/RNA sequences, including his construction cost of the Trex. This project was DNA research into animal species related to raptors' ancestors. Allison assumed it was another research project. If she had read the document, she would have realized it did two things. First, it allowed Isaac complete discretion over how the money was spent, and second, it would eliminate the management charge upon Isaac's resignation or retirement. Allison would look excellent and stand out as a person of the year due to this reduction in management fees.

In addition, when Isaac finds a corrupt company in an acquisition, he freezes the company's assets. Generally, significant cash assets were never reported because these companies were laundering money. Since there was no record of it, he would funnel about half into his pet project, and the other half would be used to pay to satisfy the investors. Isaac also channeled some of that money to fund Mantle Works' scholarships, making Allison happy. Allison also helped him immensely when she discovered corruption in competitive companies, which he would exploit.

These companies were purchased for a fraction of their value, and the executives would start disappearing - permanently transferred. Typically, these involved a miraculous accident, always ending up in the recovery of the executive in a very small baggy. Isaac understood Chadoom's role in these incidents and admired his creativity. His favorite was the executive killed flying across a military test range during a live-fire joint exercise with live nukes. Somehow, the executive's ship did not register the military ranges' no-fly boundary. Even more interesting was the military's incident review, in which the executive's ship was registered as a target drone for the ships in the exercise. The exec's ship was vaporized, and nothing was left but a few metallic fragments. Isaac had wondered how Chadoom had set it up but understood precisely how talented he was. In many ways, his robot dog Twitch was a minor security measure to keep the "Big Dog" at bay should Allison ever sic him after Isaac.

Isaac and Mackena entered the cruiser's control room. Mackena turned the Hologram on, which showed the complex buried under the island. She then switched on the newly constructed tunnel that would allow Tyrex to access the universe. "When do we test fly our special ship?" he asked Mackena.

"If we get the force generators set up in two or three weeks, we've still got a problem. We must move the ship past Cirrus Major's orbit without the security force or your boss noticing. If she finds out, she'll take the ship, make us vanish for good, and then use it to build an advanced fleet." Mackena looked over at Isaac, the seriousness of their plan clear in his eyes.

"With the Trex, the new anti-ship missiles, a new set of tactics, I estimate we could single-handedly nearly take out a battleship. A dreadnaught might be another matter, but we are fast enough to escape the following situation." Isaac said.

Mackena kissed him on the cheek. "I wish you would hire Annabell as your pilot and stay out of harm's way with me in the cruiser."

"Remember, both ships have jump capacity. We must figure out a system with a backdoor set of jump coordinates. That is something for you to figure out. I like the idea of having a 'safe' haven too." Isaac suggested.

Mackena's eyes lit up, "I think I might have that, or at least in the making. We bought that large asteroid with no value because it is part of an old sedimentary basin of Cirrus Minor. You have hired a small group of miners to hollow it out. I have transferred some local teams to Asteroid 16 to install a base kit we constructed here. We also sent another group out near the time you landed today." She explained.

Isaac remembered waiting for the freighter launch. "Good thinking. I should have considered accelerating the work sooner—a sanctuary in the home station. Good idea. How loyal are the miners?" Isaac asked.

Mackena continued, "They are 100% loyal because they understand they are hollowing out their new homes. They discretely signed contracts to mine for us and will not have the high rent Cirrus Minor Guild Station charges." She replied, then added, "Mantle Works will see them as a rogue operation but not interfere unless they start impacting Cirrus Minor's ore quotas. Then, they will send part of Allison's flotilla. The hollowed-out asteroid is excellent protection. Since it will also be armed, it will take a lot of firepower to break. We have guaranteed safety as it is our home, too."

Isaac smiled. "I will be happy to rid myself of that conniving boss and have independence!"

Chapter 6
Cirrus Minor - Prom - Space Station

Many old traditions, such as prom dances, have never disappeared. Proms were rediscovered by the miners needing some socialization after months of solitude that required social interaction. There was a dance for all adults at least once a month. These were always well attended. This fit the mining culture steeped in a deeply religious culture that supported the Guild through their dedication and success. The miners believed their faith would bless them. In most regards, the miners had been richly rewarded. These highly educated rockhounds lived well and held to traditions carried on by their parents. Calling this a Prom or a Senior Dance is misleading. These students had nearly the equivalent of a master's degree or higher (21st century). Thus, the Prom or Senior Dance was carried on. The Guild supported the annual high school prom and encouraged full participation.

John spent the day moving his computer to the sled. Jemma's components fit as planned. Since Jemma had a long-life power source fueled by Californium, she (John referred to her as she) activated it immediately. Jemma designed three additional interface boards to tie into the ship, which kept John busy most of the day. The senior dance had been forgotten about for the most part until his mother walked into sled two to remind him. "John, need I remind you that you have an hour and a half to prepare for your event tonight."

Jemma's hologram had disappeared when his mom entered the sled. "Sue's mom called me a few minutes ago to let me know Sue had her braces taken off today, and her mouth is a little sore. She might be a little quieter than normal. The doc warned her about taking it easy with food because she might bite her cheek."

"What? Did you say Jaws will not be Jaws anymore? Tony will be highly disappointed." John was referring to an old spy movie where one of the villains was named Jaws due to titanium metal teeth implants.

"Yes, be understanding if you would please?" She looked around sled two and noticed the active screens and the computer in its cradles. "Looks like you are nearly ready for your summer mining season. This old sled is what your dad and I started in, and it is a solid sled. I always liked the fact that these older models have more room. Have you considered your fourth miner - you have four private berths?" She asked and then continued with a thought: "I guess I could go with the three of you." She knew that would drive her teenage son crazy.

John reacted, "Mom, who wants to mine with their mother! That makes me cringe, and it is totally creepy. Tony would throw up, and Tanya would go out of her mind. Remember Tony went with his dad last year and has not touched Curry for the past year."

"Oh, is the computer up and running?"

"Yes, Mom and I figured out a way to modify the engines that should dramatically increase the speed. Sled 2 has always been a little clunky. Can I print some parts and do it while Dad is away?" John and Jemma had figured out how to modify the engines to get nearly 100% efficiency from them. Jemma had already sent him the file that would go straight into the printer interface to print the parts, but she had not told him about a few other things yet. They would also add two additional high-performance engines to the sled. It would be faster than the Mantle Works' new generation fighter. It also involved upgrading the reactor, which John would tackle when he printed the parts. John hoped to use his graduate thesis to find diamonds within the belt. Turk kept a well-stocked maintenance bay, and John kept an inventory of what he used to pay his father back after the mining season.

"Your dad sent a cryptic message and said he would be delayed getting your uncle Adrian out of his situation. He should be back in a week. Go ahead, but I would like to see some testing after you finish installing everything. A spaceborne trial will wait for your father's return."

"Thanks, Mom. I will make sure this gets tested before we head out," John enthusiastically replied

"John, You need to get dressed, and I would like to take some halo pictures of you and your sister before the dance. "

"OK, Mom, give me five minutes to wrap up." His mother turned and left through the bay door.

Jemma's image emerged in the center of the room. "John, I have fully integrated into this ship. I have several other improvements to suggest. When would you like to discuss these?". Jemma said.

"Thanks for standing by." John realized he was working well with his new friend. "It must be tomorrow morning after we print the engine parts. I will see if Tanya is up for some classic engineering and ask her to help with the engines. She is, after all, a better engineer than I am," John answered.

"Throw me a bone, and I will chew on it." A bone materialized in Jemma's mouth, and she started to gnaw on it, making loud noises.

"Very funny, a computer with humor - this is a first. Later". He turned and started to walk out.

"Hey, don't I get a treat or something? And you are supposed to say, "You will be right back," thinking I don't have time. Stupid dog owners, maybe I should have chosen an earth skunk and smelled the place up."

John shook his head as he pushed the bottom to close the sled. "That monster is going to drive me crazy."

After a shower, John changed into his rented suit and combed his hair for once. He wore a black tie, ruffled shirt, black trousers, and patent leather shoes. He looked formal and sharp. When he entered the living room, he saw his dazzling twin sister in a green strapless dress accentuating her figure. John had never noticed.

Now, he thought he needed to carry a small handheld laser to protect his sister from the hand of any young man with ill intentions as if Tanya needed protection. She had a black belt in Aikido. She had tortured several young men in her last tournament, one of whom had been a bully to the twins when they were young. She did not prolong his punishment because she knocked him out in three seconds of the first round in the championship. John realized at that moment that she could be tough in one moment and win your heart in the next.

"Wow, Tanya. I am glad Dad is not here. I am afraid he will start your dates in the future with a meeting of the minds. Remember Great Great Grandma Joan's journal where she described her father needing to meet his daughter's dates with a shotgun. I had to look up what a shotgun was, and it would have scared me silly. Tanya, you look stellar. Hey, tomorrow I need your help. It seems there is a dog that has some heavy-duty plans that I cannot do as well as you."

"Sounds like a golden, sure," Tanya said.

Tony came through the door in his black tie suit and stopped cold. "Wow, I had no idea the two of you could look so amazing. Mum must have snuck some curry into my lunch, and I am hallucinating."

John and Tanya's mom walked into the living room. "Well, look at the three of you. I must take some halos and send them to your Dad and Uncle Adrian. They won't know what to say. "

Tony was quick to the draw. "How about where are my children? Or leave them alone with their mothers and lose years of proper training!"

John replied, "Tony, it is not that bad. We only need to do this for three hours, and then it is back to the jumpers."

Tanya added, "Come on, lads, we need to branch out sometime. Embrace it and man up, boys!"

The three endured the halos and the numerous poses the twin's mother required.

"Come on, John, I need to pick up my date and get metal-mouth," Tony announced.

"Tony," John's mom said in a corrective tone, "Sue…."

John interrupted, "Sue is ready. Let's go, Tony," and he hustled him out the door. John wanted to see Tony's face when he saw Sue.

Tony walked out with derogatory comments about the metal apparatus on Sue's teeth. Before they picked up Tony's date, he stopped in the hallway and said to John. 'John, whatever you do tonight, do not kiss Sue - you know, tongue wrap style."

"What?" John questioned.

"I heard of a guy who nearly lost his tongue because he caught it on his girlfriend's braces and nearly pulled it off. He ended up in the hospital with deep cuts on his tongue and nearly bled to death."

John rolled his eyes. "Tony, it's not going to happen. I am a nice young man and respect Sue, and she is so damn smart. I have not met anyone who can handle astrophysics like she does."

They picked up Tony's date and walked down to Sue's home. Tanya stayed at home, waiting for her date to pick her up. Sue's mom opened the door and invited the two youths in. "Sue will be here a moment," Sue's mom announced. Within seconds, Sue rounded the corner.

"Well, what do you think?" Sue said as she rounded the corner, speaking to her mom. "Oh, I did not know you were here already," she said, embarrassed and shyly smiling.

Both young men stopped in their tracks. John was amazed at how beautiful the young lady standing before him was in her yellow dress with short sleeves. Her dark hair had large curls with diamond-like glitter. Her smile was almost as mesmerizing as Draconia's. John was taken aback. On the other hand, Tony was gob-smacked and grabbed John's elbow due to his instantly weakened knees. All he could say was, "Kiss away, my friend."

The dance was fantastic. They kept dancing until the senior party wrapped up and then headed to Tony's place. There, they played board games like Galaxy Risk and Mine the Asteroids until the early hours. Afterward, John walked Sue home, along with Tony and his date. When they said goodnight, John gave Sue a kiss on the cheek, much to Tony's chagrin. John didn't return home until 3:00 AM to find his mom and sister still awake, waiting for him.

John was surprised they were both up but found Tanya had also gotten home roughly one-half hour before them. "Wow, morning, mom, late night?"

"I thought I would wait up for my beautiful twins. Events like this only occur once in our lives. My two babies are all grown up and going out on a dance. This is one for my journal". The twins knew their mother kept a detailed journal of their lives filled with halo pics and videos. They also knew these would be used to embarrass them at events such as their weddings. Every time their mom could, she would drag out the journal, expose tidbits of videos of the twins, and swap stories with her friends.

"Thanks for waiting up for me, Mom. I know you were doing that to see if I ended up with any lipstick on my face," John spouted.

Tanya came over, planted a big wet kiss on his cheek, stood back, and smiled. "That's alright, John. I think I just covered any evidence because I have your back. That is what twins do, and they always know what the other is thinking and going to do next."

Chapter 7
Cirrus Major: Spies and Football - Allison's Beach Cove Mansion

Technology is the key to growth, even in the sporting world. Time was a harsh mistress to the humans caught in the sporting world. The traditional sporting world has morphed from the conventional set of sports in the twenty-first century Earth to a couple of solar systems. Traditional American football morphed from large professional organizations with massive taxpayer-funded stadiums to highly sophisticated computer programs linked to avatars and two opposing human stars who could play any team position. Each planet was organized into geographically based football leagues that played most of the year. The better players with the best technology rose to the top but were split into two groups. The athletes who thrived have substantial fan bases, are reclusive, and shy away from public contact (Remember, the fans only see the avatars). Chadoom Clements was one of the latter who loved technology and was a natural athlete who strove to ensure the game was fair and honest. He carried a large fan base even though he was a recluse to them.

Chadoom was in the halo video immersion room, a high-tech space he had set up five years ago. This immersive environment allowed him to fully experience Man-Up Football. This action sport evolved from North America's traditional National Football League (NFL). The modern version emerged as the NFL's popularity waned, primarily due to its players becoming overly involved in various political causes and focusing predominantly on issues of certain minority groups. This shift alienated many of their core fans.

Additionally, the players' skyrocketing salaries, fueled by greed, had led to unsustainable financial demands on the sport. This escalating situation reached a critical point in 2205 when the NFL player union initiated a strike following the preseason games, signaling a significant shift in the landscape of American football. Chadoom's engagement in Man-Up Football represented his passion for the sport and embracing its more inclusive and sustainable evolution.

The strike lasted a year and saw the collapse of ten of the thirty-five teams. The union did not live up to its promises of paying union members to offset funds, and the union leaders were investigated for fraud. All were convicted but were pardoned by a far left-leaning socialist North American president. The players finally realized they were losing their jobs and careers, caving into the remaining owner's new demands of a massive salary reduction. The fan base had lost interest because the issues the players demanded other than more money were unrelated to sports. They were all political. Their message was loud and clear, and the fans turned on them. Those fans, however, were still hungry for the primordial need to take a side.

The technological solution to the challenges faced by traditional football was a groundbreaking video game featuring teams of animated players. These players were controlled by human participants who wore nano-censored arm and leg bands, enabling their physical movements to be accurately replicated in a virtual world. Participants would line up for plays, run, dodge, and tackle, all within this digital realm. The animation technology, perfected in the 21st century through various video games, provided an exceptionally lifelike and immersive experience.

The American Football Organization (AFO) capitalized on this technology, forming city-based teams that competed in virtual tournaments. These events were broadcast in a style reminiscent of the old NFL games, capturing the essence of traditional football while embracing the new virtual format. A significant advantage of the AFO system was its flexibility; unlike traditional football, these games were not bound by a specific season and could be played anytime. This innovation solved the issues of player demands and political controversies and breathed new life into the sport, making it more accessible and engaging for a broader audience.

The groups were organized in small for-profit companies on a city-by-city basis. Eventually, these companies evolved into large multilevel organizations, surprisingly funded by the cities where they resided. That differed from a town supporting an owner and their team with a stadium. The actual players were never pictured or highlighted. Most of these players were geeks or nerds and would have paled against their graphic avatars. Fans would have run away if they had seen the appalling physical shape of most of these couch-potato athletes. Visual development and its associated research were their highest priority, which paid off. People gathered behind their city teams and could watch from their homes, pubs, and sports bars. AFO made access accessible to all citizens. Cities promoted players' avatars who became heroes in the community. The competition was tough.

The evolution of the game's graphics and interface was remarkable. It transitioned from using arm and leg bands to a compact personal arena, where advanced 3-D computer graphics accurately mirrored players' movements. Due to space constraints for hosting an entire team in these new arenas, the game's format was refined to include only vital positions. Thus, staffing was effectively reduced. The culmination of these developments was the introduction of Man-up Global Football, a fully immersive game that represented a significant breakthrough in the market.

An artificial gravity arena made the game very realistic, with the player in the arena subjected to genuine impacts through variations in the arena's gravity fields. Man-up replaced the forty-four-man teams and the old nano-sensor suits—a single individual who would switch roles for any play. The rules were much looser, and the player would sustain injuries if the football players they represented were injured. Only the fittest would survive, and true athletes emerged.

The traditional scoring system in the game was enhanced with an innovative twist: off-field judges awarded additional points for unconventional plays like cheap shots, using deflated balls, and ugly hits, provided these actions avoided swearing and obscene gestures. However, if a player was penalized for any such move (indicating they were caught), the potential extra points for that action were nullified. These additional points, evaluated by the judges, were tallied and added to the team's total score at the end of the contest, adding a new strategic layer to the game.

Chadoom had risen to the top of the Cirrus Majors League, but he had always won based on his athleticism, unlike his competitors. He disliked cheap shots and hits designed to inflict damage. The arena was an octagonal-shaped room with a floor that was an omnidirectional treadmill. The floor was Chadoom's arena. Chadoom used a 3-D immersion helmet colored bright orange with an emblem of a horse rearing up. The orange helmet matched a Lycra-coloured orange suit with blue piping he wore. He was exceptionally fit, and the suit molded around him like a glove. Again, his football team was avatars dressed in the same manner. Each character was a computer simulation running on very sophisticated programs. Today was a big game that Chadoom had been looking forward to.

The audience had gathered in homes or pubs hours ago, going through various pregame traditions. Those gathered at pubs had plenty of time to launch themselves in an inebriated state if that was their choice. The game had a huge following, and they measured the audience in tens of millions, making it a marketer's dream. The advertiser had only two commercials allowed during the four-period game. This drove the price of advertising to highly competitive levels. The players, by default, received a small portion of the funds from advertising.

Chadoom's fortune had already soared high before his football ventures began. He was so affluent that buying a small, inhabited asteroid was nearly within his grasp. Yet, his unwavering loyalty to his friend, the CEO of Mantle Works, and his passion for his regular job held back any thoughts of early retirement. Now, at the championship game's halftime, Chadoom was lagging by two touchdowns. He used this time to rethink and craft a few new counter-strategies. An interesting fact lingered in his mind: his opponent had constructed a transparent arena, offering tickets to spectators to observe the physical aspects of the game.

On the other hand, Chadoom played in high security at Mile High Stadium as the Orange Crush. His opponent had frustrated him in the first half. It seemed his opponent knew Chadoom's playbook. The third period was about to start. Chadoom suspected something was off and had set up a special firewall. He had preprogrammed his first set of downs during halftime. He was looking for evidence of tampering because his opposition knew his plays before he executed them. This was beyond chance! Chadoom played the role of the kicker on the first play. After kicking the ball, he prepared for what he guessed was coming: a cheap double shot from two blockers. The kick receiver had received the ball and ran it out of the end zone. This was the distraction as the actual play, Chadoom guessed, trying to take out Chadoom with an injury or, better, a knockout.

He knew two players would be allocated, with one being his opponent and a hitman. One would be a tackler hitting him high and the other low. As he was running down the field, he had time to spot the two players designated as an assassination squad. One was his opponent, but he did not know which one. Someone was feeding the opposition inside information that could only have come from the graphic interface technician. He had seen a telltale signal sent from his team's technicians. They would have his plan in hand. He would deal with that individual later. Opposition players number forty-four and twenty-seven both cut precisely toward him at the same time. They committed with twenty-seven diving low and forty-four high. One would be a low-cut block designed to take the knees out. The other, hitting him high from the opposite direction, would attempt a helmet pile drive to reverse the cut block's momentum to maximize damage.

Chadoom was prepared, however, and jumped when the cut blocker committed his low block. He then stiffed-armed the high hit, which changed his momentum, and he pulled his legs to his chin, landing on the ground with a roll in time to be smashed with a vicious elbow. Chadoom jumped up, saw the runner with the ball, and tackled him. The runner dropped, and the play was over. Chadoom turned in time to see the telltale sign of his opponent leaving the tackler committed to hitting him high, who is now lying on the ground due to a quick shift of eye color.

Chadoom knew the opponent would be the running back or the quarterback on the next play. He would choose the quarterback next. At halftime, Chadoom also sent a small robot to his opponent's public arena with the potential to execute a special mission. The teams lined up again. Chadoom took the position of a blitzing right linebacker. He needed to break through the line to go after the quarterback, which he guessed was a pass play. He accessed a private channel to his small robot, which was shaped like a fly, to execute its program and momentarily disrupt the magnetic field generator by tripling the gravity on a single pad in the line Chadoom would hit the quarterback. This would create an invisible gravity load that triples the weight of anything in that pad. Chadoom's target was to hit the quarterback so that the quarterback's head and shoulders would enter the overloaded gravity plate. He had to ensure he did not fall prey to this extra gravity force through his momentum. Chadoom's arena was a mimic of his opponents. His goal was to knock his opponent straight out of the game as his opponent had just tried to do to him. Chadoom's little fly under the gravity field had one shot to alter the gravity field due to the energy required.

The center hiked the ball. Immediately as Chadoom charged the line, he could see it was a pass play as he had predicted. He had a straight shot on the quarterback and pressed the issue. He lunged at the quarterback, hitting him at an oblique angle, which drove him back and down. The contact also knocked the football loose. Chadoom's adrenaline surged, and his world went into slow motion. First, he saw when the quarterback entered the disrupted gravity pad. His head and shoulder tripled their fall rate, and he heard the crunch of at least a collarbone, if not more. The football popped out of the quarterback's hand and bounced hard outside the impacted gravity plate. Chadoom could be caught and penalized if it entered the abnormal gravity field.

Fortune smiled at Chadoom as the ball unexpectedly bounced his way. Stretching out, he scooped it up from the ground in a swift motion. With an open field ahead, he dashed towards the touchdown. Chadoom was aware that his agility had given him an edge; his opponent, who could have caught up, was now out of the picture. Though not the fastest runner, Chadoom's quick thinking turned the tide. The crowd erupted excitedly, cheering for his impressive eighty-eight-yard fumble return for a touchdown.

Only after a small touchdown celebration did the crowd realize the opposition quarterback had not moved. Like the actual game of old, the opposing quarterback needed to be carted off the field on an air sled. In reality, an emergency medical unit entered his opponent's play arena and scoped him off in an air ambulance. The helmet and suits had detectors that transmitted the player's health. The medical team knew immediately their patient had a fractured skull, concussive brain trauma, and a compound fracture of the collar bone. It wasn't pleasant. The little fly performed its final function, flew away from the venue, and self-destructed in the local drainage ditch.

Chadoom readily won the game by four touchdowns. Chadoom walked out of the arena, thinking one thought. He had made deflate-gate look like child's play. They would spend months tearing apart their opponent's arena only to conclude the gravity square malfunctioned but not know why. His opponent would play next year but would take a while to recover. Chadoom now had a small job to handle. He had been targeted by corruption inside his own team's staff. Someone had betrayed him to his opponent, and he would find out who. It would have been good advice for his technician not to upset a brilliant serial maniac.

Allison met him outside his arena. Chadoom's house was built on Allison's expansive one-hundred-square-kilometer estate. Allison had come over on her hoverboard, which she often used to move around the extensive grounds. Only Allison was allowed in Chadoom's arena unless there was a medical emergency, and those people were Allison's estate medical personnel, which Allison had upgraded due to Chadoom's football interests.

"Nice come back, Chadoom! I thought the first play of the second half was exciting and a touchdown that seemed to turn the game. It sucked the life out of that team. Did O'Leary survive that hit you put on him? - looked like a glancing hit, but he went down hard." She said, smiling as she stepped onto his veranda. "Get a shower; we have a business to discuss, and I will treat you to dinner, you very clever man. We had an anomaly reported today of a small electrical anomaly moving across and out of the compound sometime near the game's halftime. I let it go because it did not seem a threat. It would help if you looked into it, however. You know how devilish small robotic probes can be."

Chadoom smiled. "I will get my shower and meet you there in half an hour. Anomaly, you say? I am sure it was nothing of importance. You can count on that."

"Be back in 30."

Allison went to Chadoom's fruit bar and extracted her favorite, Strawberry/peach, which seemed to fit her mood. She sat, reviewed her messages, and ordered her aircar. Chadoom was ready to go to dinner one-half hour later, and they hopped in the aircar.

"Darn network, want to interview me tonight on their Man-up highlight show. I hate putting on my football disguise to do a 15-minute interview." Chadoom wore a disguise consisting of a mustache, wig, and a couple of cheek lifters to prevent easy recognition. He carried it in a small bag and took no more than ten minutes to put on when needed. He would put it on tonight after dinner with Allison when they finished their business.

"That's is part of your popularity, mister football hero. The kids love you, and your youth foundation is flourishing. You are doing good work and are looked at as a role model. The things we hide in our lives. Mister Perfect is on the outside, and the serial maniac is on the inside. Chadoom, my dear, we are just two peas in the same pod." Allison stated that when they got in the aircar, "Now, let's go back to that anomaly and the first play of the second period. How are they related? I bet, knowing you, that you somehow modified the playing field. I also read the research report on miniature probes you had your spy nerds create, and I know they have multiple uses." Allison rarely missed much, and Chadoom knew where she was going. He had nothing to hide from her. She continued, "Perhaps you have continued to use your very creative skills to sway circumstances in your favor in that game. Well done!"

"You are correct, of course. One of my video feed controllers provided play information to O'Leary in real-time. He knew my position and play-calls every play in the first half. At the halftime break, I downloaded the transmissions during the first half and tagged three suspicious links. In the first kickoff of the second half, he targeted me to take me out of the gameplay. Since I knew where to look for it, I dodged it. I had sent O'Leary a little friend in case I could catch him in the act. That is what you recorded exiting the compound. It's a good test for our system, however. The little friend I sent had a single mission to help me; let's say I cleaned up a little social injustice."

Chadoom continued, "O'Leary has a fractured collar bone and a severe concussion. A fractured collarbone might be an improper description of the damage. The surgeons are arguing about replacing the collarbone with a synthetic or putting a plate with screws galore. He should be out for about a year. I will have an honest chat about bribing the controller at some point. Of course, the controller will be replaced. I hope we can hire him into Mantle Works to create a sting and send him to the island with your permission".

"Hmm, Chadoom, you are so creative. Keep him out of confidential areas. I realize he will be targeted and provided a relocation to the island with the risk that he might not fall for the sting you set up."

"If he does not fall into the trap, he will have a potentially bright career. I can live with that. But he won't because the temptation will be too great."

"Fine, I wanted to discuss using your little bug friends with you. I was reviewing some financials around some of our recent acquisitions. There are some potential discrepancies, but I can't put my finger on them exactly. Some funds have been diverted, but all are within the company. I don't know where they have gone to. I don't understand; the issue seems to be pointed at Isaac. I trust Isaac entirely for what he has done for the company and his dedication, but I would like you to drop some of your friends his way. He was playing golf this weekend at his favorite resort."

"I can handle that. I will set up his office and the golf resort this week." Chadoom replied.

"Good of you. Thank you. Now you can tell me exactly what your little friend did to poor O'Leary and where it is now."

Chapter 8
Cirrus Minor - Extraction – Cirrus Minor Asteroid Belt

Mining sleds are impressive vehicles. They are spacious and built to survive. They included enough for a family or a small party of miners to live comfortably for long periods. They were also well equipped with plenty of technology. Most miners loaded their sleds with their own design specialty tools that worked with their personalities and mining styles. Different mining goals required different load-outs. Most sleds towed a storage bin or two for tools, supplies, and, most importantly, recovered materials. The miners were not only innovative, but they were also creative. One of the most interesting on the creative side was Adrian. He had a degree not only in mechanics but also in inorganic chemistry. He built some unique tools that Turk had only seen a fraction of. However, Turk solved problems like extracting stuck mining sleds with his level head.

Turk climbed into Adrian's sled. When he removed his helmet, Turk was relieved that he didn't encounter any overpowering curry odors. The mentholated Vaseline had been a wise choice. Then, he noticed two diamond crystals next to the sample bag. Turk exclaimed in surprise as he picked up the larger diamond, "Wow, Adrian, are you serious? Where did you get these from?"

"Exactly what we've staked our claim on," Adrian declared, emptying the sample bag. He carefully unwrapped the core plug extracted earlier from the chamber. "Found some diamond-bearing kimberlite here. I'm willing to bet this rock's hiding more."

Turk was now mulling over the pink diamond surrounded by tiny crystals and diamond shards from the sample bag. "Adrian, I'm speechless. I thought you had found some minor platinum or palladium on this rock. Why are you sharing this with me? Just this," he held out the pink diamond, "would keep you out of the Asteroids for a year with a new sled. Home-time, my friend, is what we all wish we had more of."

Adrian broke into a smile. "My good friend Turk. Our children are joined at the hip. We have been friends forever. You have saved me from situations like this. How many times? How often do you find a friend you can count on anytime? You are a ministering angel to me, and I love you like a brother. And families are forever. So this will improve both of our lives, and I will certainly be able to tinker in the hangar more. How could I not include you in this good fortune?" Adrian grabbed the mining drone and pushed the button to connect it remotely. "I have not looked at this in detail. What do you think about doing that now?"

Turk nodded in agreement, his actions speaking louder than words, as he was a man of few. Adrian dimmed the lights and said, "I thought the small diamond would be a fitting introduction. Brace yourself; I believe this is going to amaze you!"

Adrian switched the Hologram projector, which formed its screen. The 3-D picture was looking down into the round hole. The drone moved up inside the cavity through the side of the cut into the cavity, revealing the contact with the eclogite. The bottom of the natural pocket showed diamond crystals posed proudly above the background mass of diamond crystal shards. At the bottom of the image, Adrian's faceplate moved into view as he followed the drone. As the drone dropped, the lateral extent of the cavity was lit. Light refracted from the diamond crystals lining the cavern. Reflected light shards were coming from all angles.

The hologram displayed a vertical and lateral scale, allowing for the measurement of distances and individual features. As the cavern expanded from the opening, it revealed diamond crystals that reached up to eight inches in size, each in perfect octahedral shape. The sight was truly astonishing. While Adrian had previously viewed this on his wrist monitor, he had no idea of the magnificence of the actual image. He paused the screen to fully absorb the breathtaking scene. Within the cavity, most of the crystals appeared as single entities or clusters of interpenetrated octahedrons delicately arranged on the surface. Other crystal forms could also be seen, including dodecahedrons, cubes, and tetrahedrons.

Turk broke the silence, "This survived a planetary collapse. I could not even dream of such a thing. Can you switch to a spectral display?"

"Yes, I had an infrared spectral analysis as part of the probe configuration. Diamonds look to be dominated by type 1a with a nitrogen peak at 1292 angstroms and low boron. A chart was extracted in 3D space. The halo switched back to visible light and slowly moved into the cavity. The probe is half a meter from the base, and the cavity is about two and one-half meters wide. The same brilliant scene of high-intensity light shards reflecting from the crystal facies played out. "The probe is one meter into the cavern."

"The question is, do we take the vug out whole or mine it in place?" Turk wondered.

"The answer may be what comes next," Adrian answered.

The probe moved another meter into the vug. Interference briefly impacted the picture. Finally, a bright, iridescent flash caught the camera, which meant only one thing - Draconia. A single crystal was prominently sitting on a large pale green diamond at the edge of the field of view.

Turk exclaimed, "Look at the Draconia in situ. It has never been reported in situ in Cirrus Minor. Draconia on the matrix is rare. This will add significantly to the scientific community's understanding. My good friend, you may have discovered one of the largest Draconia crystals ever. The fact that you used a probe to investigate this was smart. You should save it and copyright it. It's worth a lot of money."

"I was following an old copy of Turner's map, which led me here. Draconia in Cirrus Minor will be noteworthy. I was following a lead he noted on the map. He found the biggest deposit of Diamonds ever found but never mined it. He said he would sit and stare at it for days. He said he could not mine it because he would destroy its beauty. I see why he called it 'my precious,'" Adrian stated.

Turk chuckled and remarked, " 'My precious' – that phrase is a classic from a story by a man named Tolkien, set in a fantasy world called Middle Earth. The main character was attached to a magical ring, not a diamond vug. It's unlikely anyone would write a fantasy tale about a diamond vug."

The end of the cavity was now becoming visible, resembling a chandelier bathed in sunlight within a dark room. The hologram dazzled with reflected light, capturing the iridescent flashes from both the large and smaller subordinate Draconia crystals. It was a mesmerizing sight.

Adrian couldn't contain his excitement as he exclaimed, "The diamond crystals are perched on a finer mass of diamond, which transitions into cluster masses. The clarity of these diamond crystals is exceptional. I believe we're witnessing a variety of pastel-colored diamonds, much like your pink one and the green diamond the Draconia crystal is resting on. This is so exciting I may pee my pants," Adrian exclaimed.

"This vug is five and a bit meters long, two and one-half wide, and a meter high. I think we can cut it out as a single piece. The most challenging cut will be the roof cut, where we drop the vug into a net."

"I have a chemical bot cutter that might do the trick. Stuff goes everywhere. Hydrofluoric acid cutting is tricky, Turk. How about we cut my sled out by cutting under it? Then we can let it drop away from the pipe?"

"Could do that, but it is risky to the sled? We might, however, make four cuts under the sled and winch the sled out. The cuts would create columns until we use the chembots to shear the bottom of the cuts off. The eclogite should brecciate and act like rollers. That would keep the cavity safe and minimize damage to your sled. Probably best if we detach my main plasma cutter and mount it on its gravity sled."

Adrian switched the halo model to his ship wedged onto the large crevasse. He exaggerated the scale to look at clearances. "Might work, but we should build a trench here to let the eclogite rollers or sled drop away. A little surfactant pumped under the ship might also help with the sled action."

"Never thought of pumping surfactant in a vacuum. It should be easy to distribute where we want it," Turk replied. "My winch is rated 150-kilo tons, which should provide enough leverage. Your sled is the version marked 27, right?"

"Well, 28, no difference. Twenty-five kilo-tons loaded minus the plasma torch, which is that pile of rubble over there," pointing to the lump of debris in the halo stream. "Does not work well when scraped off the top of a ship," Adrian quipped.

Turk frowned." I'm not the crazy spaceman who drives his sled into a crack in a meteorite because some dude's map said to! "

"Well, I did take it slow and look at the lagniappe." Adrian picked up the large diamond crystal and gently tossed it toward Turk.

"Whoa, a gift. Thanks, pal." He tried to stuff it in his vest pocket, but it would not fit. "Here, he returned it by handing it to Adrian. "Too big for my pocket. You better hold on to it."

"Ok, we cut under the sled. Dig a hole in front of it and pull hard. Right?" Adrian summarized. "I will get a blade out and smooth a flat spot to set up the plasma cannon from your ship. First cuts can be cut from your sled. Do you need help moving the cannon once you have completed the cuts on the debris hole?"

"No, keep your head down on your blade. I will signal you." Turk said.

They both suited up in their sturdy mining shells, evacuated the air, and opened the rear of Adrian's sled. Turk headed into his ship to access the external plasma cannon controls and retrieve the quick-release tools he needed. Meanwhile, Adrian emerged, steering a compact vehicle known as a 'blade' that he had stored in his storage compartment. The blade unfolded, revealing a robust blade at the front and a set of mining claws at the rear. Positioned on the side was a heavy rock drill, currently set at a 90-degree angle. Then, Adrian brought out a small trailer filled with twenty-five three-centimeter tubes.

From Turk's perspective, it looked like an old military high explosive rocket launcher and ran on track treads, he thought, as he aimed the plasma cannon. A bright green plasma beam hit the ground opposite where Adrian was working and slashed parallel lines in the ground. Turk was digging the landing zone for the sled when they pulled it out. As the plasma hit the rock, the plasma backscatter nearly blinded as Turk dug the one-meter-deep trenches. The plasma beam was incredibly fast. It cut the Eclogite like butter. Adrian drove the tractor around and waited on the side until Turk stopped his cutting.

Once the cutting had stopped, Adrian detached the trailer and drove the blade perpendicular to the still-glowing trench. He dropped the claw into the top of the first trench and used his hydraulic leverage to rip the Eclogite between the parallel grooves. The rock brecciated as he drove slowly across, ripping the surface to about half a meter at a swath. His first pass was rough as he pulled out Eclogite blocks formed like columns. He then drove the middle of the patch with the claws extended, ripping the blocks of rock into pieces. Adrian turned to follow the line until he was at the end of the cuts, turning again to rip perpendicular to the grooves. He followed a Zamboni pattern when it cleaned ice in the 20th century in hockey arenas from outside to the middle in decreasing ovals.

Once he had the material ripped into smaller pieces about fist size, he dropped the blade and plowed the debris while pulling another half meter down. Then, he cleaned the waste out through the exit ramp. No additional kimberlite pipes were exposed. When he bottomed out of the precut grooves, he bladed a flat spot at the base of the exit ramp.

Turk was waiting up top with the tractor, pulling the plasma cannon mounted on a utility sled. He drove the rig down the ramp to the flat and lowered it to the pad. Adrian grabbed the power lead and cable and released it to his external power feed. Turk efficiently lined up the laser for the first cut in short order. "Adrian, my only concern is where your ship's base rests. Let's take a quick sonic survey to know exactly where your ship's base is."

"Ok," Adrian said. He pulled out two mining drones, one labeled thumper and the other listener. These were typically used to test for layers or density changes in an orebody. Adrian placed them into a hatch that accessed the outer shell of the double-hulled sled bottom. The drone-labeled listener split into six components and formed a star pattern around the thumper, spaced a minimum of one meter away. The drone labeled thumper landed on the inside of the outer haul and started to vibrate, beginning at 1 hertz and slowly increasing to 120 hertz. Once complete, the drones all moved one meter down the hull.

Adrian watched a picture develop in 3D showing the interface of the hull and the rock where they were in contact. Those parts that were not in communication with the rock only showed the hull. Sonic scans required something to pass the vibrations, and the dead of space provided no medium to transmit the vibrations through. The drones made short orders for the ship's survey. Adrian opened the hatch and directed the drones with his wrist computer to tie the study inside the vessel to the pit floor for proper related elevation reference. Once complete, he directed the bots back to their sled storage slots. He sent the signal to Turk's wrist computer, then fed it to the targeting computer's plasma cannons.

"OK, Adrian, I am set here. Are you ready?" Turk said as he fine-tuned the cannon for the first cut.

"Yes, please don't cut my sled in half. I have a lot of precious memories tied to that ship. I will set the chemical cutters to a height of .15 m off the floor of the cuts. The chemical bots were an interesting device. When launched, they looked like small crabs and would crawl along the bottom of the cuts until they reached their programmed location and climbed the wall to the preprogrammed height. They were mostly used in cavern mining to collapse walls or stoping roof blocks. Adrian would cut holes and set the bots to work while maintaining a safe distance. The bot was a two-stage device. The first stage was a shaped explosive charge, and the second was a highly corrosive hydrofluoric acid bath released due to the blast just after the shaped charge was detonated. The acid would cut through anything the horizontal shape charge did not. The result would be a horizontal cut through the Ecolgite perpendicular to the vertical cuts Turk was starting. If the plan worked, the sled mass could free the ship, even in the low gravity field."

"Ok, Adrian, I made the two outside cuts. I will do the first center cut and then work away from it if the ship's weight crushes the vertical cuts." Turk told Adrian his cutting plan. Adrian had set the program for the bots on his wrist computer and transferred the programmed information into a short wand, which he pulled out of his utility belt. "OK, I will be with you in a moment. He waved the wand over the top of the trailer, holding the chemical bots working length-wise to program them. He then connected the blade tractor to the sled, drove it down the ramp, and parked behind the laser cannon at a distance. It was about safety, and it was known that plasma and explosives did not mix. "Turk, you did not trigger any alarms on the sled during the center cut. That is good. I still have a whole sled."

Turk said. "Lucky you. I set the maximum depth from the survey to one centimeter below the hull. Hey, those are slick bots. You'll have to make some of those or give me your plans."

"Plans? Do I have plans? It should still be in the 3-D printer memory. They do a good job. Cuts look good, and I won't tell you where I stored my curry powder."

Turk laughed while realigning his cutting program. "I will be done in a couple of minutes, and I will then attach the cannon to my sled and grab the winch cable," Turk announced"

Adrian returned, "I will let the hounds loose."

"What hounds?" Turk questioned

"It's an old earth expression and a song Tony taught me in his recent school project about ancient songs, 'Who let the dogs out? Woof Woof Woof Woof!'" Adrian sang out.

Turk shook his head in bewilderment as he cut, "OK. I am done. Let me lock the cannon. And I will drive it up to my sled and move the plasma cannon back onto it."

Adrian still humming the catchy little tune his son had taught him. "I will let the rest of these out and see what they do for us."

Adrian hopped on his tractor, to which he had reattached the trailer, and towed it to the pad Turk had just pulled off. He backed the sled so the trailer's tail faced the cuts Turk had just finished. "Ok, boys, off you go."

Adrian opened his wrist computer and pushed the button he had set up when he programmed the chembots. The tiny robots poured out the tubes and crawled into the vertical cuts. Adrian moved the Blade to his sled and parked it in its bay. He took the trailer and restored that in the sled, too. "Turk, I am going to fire up the chembots. They are in place."

"OK, give me a minute. I would like to see this," Turk replied from the top of his sled, where he was reattaching the Plasma Cannon. He had used his small sled crane to lift the cannon. He finished with the last bolt and jumped off. His descent was slow due to limited gravity. "Ok, Adrian, I am at the edge of our work pit."

"Ok, be with you in a moment," Adrian said, bounding over to where Turk was standing.

He arrived, opened his wrist computer, and pushed the big red button on his touch screen. They felt a subtle bump under their feet and saw the gas released from the highly concentrated acid leaching into the fractures created by the shaped charges. The mist hung in the pit due to the lack of atmosphere. Then, the front portion of the cut under the sled crumbled and collapsed into the pit.

Turk noted, "The chembots did their work. The front of the cuts collapsed, but I don't think the sled is free."

"Yeah, I will get the tow cable and attach it." Said Adrian, turning and hopping to Turk's sled. He grabbed the lead, but the winch was neutral and pulled the cable to the back of his sled. He pulled a harness from a storage compartment and attached it to his sled. Then he attached it to the main cable and returned to where Turk stood. "Were ready to heave-ho. It's time for a little pull, Turk. Don't rip my baby in half, Please."

"That would be doing you a favor with those diamonds you have already pulled out. You would not have to worry about getting rid of that Mark 28 for scrap when you buy a new sled - cash." Turk replied

Turk opened his wrist computer, switched to his winch controls, and started it up. The cable tightened, but the sled remained fixed. Turk applied more tension. The rock in the cut crumbled, and Adrian's sled slowly slid out into the pit on a bed of stones, acting like little rollers. Turk continued to pull until the back of the sled came up the ramp and stopped half a meter from them.

Adrian jumped up and down, "I am so happy I could kiss a dog! Turk, are you ready to figure out how to get that vug out? I figure if we cut to parallel cuts with the cannon to cut the sides. Let's make them deep to use the chembots to create a pyramid cut above the chamber. It will take a while. Before we do that, let's cut the end with hand plasma torches." Adrian said.

Turk: "OK, sounds good. We will need to build a little scaffolding with a cradle to drop the chamber when it comes out". He thought he loved mining in space and how so many complexities were solved with simple solutions, ingenuity, and creative uses of technology. It was also their first step to becoming very wealthy miners.

The two put the cable and harness away. They made sure Adrian's Mining Sled was stable and then started to drag out scaffolding from their ships and put it up with its quick connectors. The rigging was ready in an hour. Adrian dropped down and reloaded the tubes on the sled. He did not add the hydrofluoric acid to the bots this time but used water. He then programmed the bots. He would split the bots into lines that would move into the parallel cuts along the side and orient at 50 degrees on each side to form a teepee cutaway from the vug. Turk finished safety checking the scaffolding and jumped down to make the deep cuts on the sides of the cavity. Adrian pulled the sled near the structure and returned to join Turk on his sled. He aimed the main ship's plasma cutter with its range finder along a line shown on the digital wire framed diagram that depicted the vug in a display with cutting lines. He then took the first cut. "Hmm, that looked pretty good to me, Adrian. What do you think?"

"Looked spot on and deep enough. You didn't even jiggle". Adrian replied. Turk lined up the second cut and made it. "All done on this end," said Turk. He opened a compartment on his left side, extracted his hand plasma cutter, and checked the fuel load. He looked up. "Good to go."

Adrian bounced to his trailer, grabbed his plasma torch cutter, and climbed the scaffolding to opposite sides of the cut. They spent another 30 minutes connecting the Plasma cannon cuts, flanking the diamond-bearing cavity. When the iridescent green flaring stopped, they silently went to the trailer. Adrian then programmed additional chembots, opened his wrist computer, and held his wrist to Turk with the red button pulsing, indicating the bots were ready with programming. Turk touched the button and said, "Best friends forever!" The bots poured to the ground and climbed the rock wall to line up neat rows around the four cuts.

Adrian held up one finger, saying," Round 1." The first line of bots hanging on the rock upside down marched forward into the four cuts. The bots needed to cut three meters of width with the cavern one to two and a half meters wide. Adrian's wrist computer turned green when the first group was in position. He pushed the button on his wrist computer screen, causing the chembots to detonate, which was evident when the second stage went off with debris flushed out of the plasma cutter lines. He sent the second group of bots in and waited for them to get in position. "We made roughly half a meter of progress." He again detonated the bots with a similar result. He repeated the process two more times.

"Turk, I have more bots, but I think we should get a pry bar in the cut to see if it is being wedged by debris from the chembot work. Several pry bars are on my sled's right-hand compartment starboard side. Meanwhile, I will send a single bot to test the depth of the cut."

"Be back in a minute, Turk replied," as he bounded to the back of Adrian's sled. Adrian sent the bot, which indicated the cut was complete, but as Adrian suspected, debris was wedging the block tight to the roof.

Turk returned with two metal bars with flat ends. They applied the pry bars on the same side of the cut and leveraged them with a rocking motion. No response. Turk pointed to the other side, and they again rocked the bars. This time, the block moved a little.

"Man, this is just like when Tony lost his first tooth. It took forever for him to have the strength to finish the job. Then, when it was nice and loose, he refused to touch it until it dropped out of his mouth. We did the tooth fairy thing and traded the tooth for a credit. The next day, he came home and pulled two more out that were not quite ready to come out. We asked him why, and he said he needed the cash for a computer game since he was getting paid for the baby teeth."

Turk laughed. "Son like Father Adrian? He sounds just like you. Thinking in directions others don't." He moved to the end of the cut and inserted his pry bar. Adrian moved to the opposite side, thinking about what Turk had said. "Hmm, I guess you are right. I do have tendencies to act on processes to accelerated results."

"Yes, but it is not a bad trait. Look at these chem-bots of yours. It would be best if you sold them to the other miners. It's a gold mine, so to speak." Turk replied as they gave a heave-ho to the pry bars. The block moved and started a slow fall into a sling the two had set up. They jumped clear for safety if the block of rock was too heavy for the sling, which could bring the scaffolding down. The block slowly dropped into the sling, which stretched to accommodate the mass limited by the reduced gravity of the asteroid. The sling held after the scaffolding bent in a little with the total weight of the block of eclogite containing the diamond-bearing cavity.

"Turk, let's put this in your sled. I would rather have this in good hands if I need to abandon my sled. I will trim the top so it will make a solid base. That way, the hole into the cavity is on top." Adrian requested.

"What, Adrian, are you finally admitting I am a better sled driver than you? I never thought I would hear the day."

"Right, I stuck my foot into that one," Adrian replied as he grabbed his plasma cutter from the sled that had contained the chem-bots. Turk took off to get the sled to transport the incredible specimen back to his ship. Adrian lit the torch, knocking down rough spots that could potentially damage his hard mining shell, and then proceeded to cut a flat base on top of the extracted block of rock.

He had just finished when Turk returned with a small gravity sled on which the stone would sit. When Adrian dropped the block, he parked it beside the sling to sit beside the sled. Adrian passed his plasma torch to Turk, who stored it for him, then grabbed the controls for the sling, slowly lowering it until it was a quarter meter from the top of the sled. Turk had dropped the sled, so it was sitting on the ground to prevent it from moving. He then moved over to help Adrian wrestle with the block. They lowered the side of the sling closest to the sled, which started to rotate when they pushed the top of the block. Adrian jumped down when it began to turn and grabbed a pry bar to ensure the block of rock would stop on the sled and not continue to roll. The block settled fine, leaving a fist-size clear diamond snagged in the sling. Turk worked it free and held it up. "Looks like we jarred a little prize loose that came out of the hole." He was thrilled by this chunky diamond crystal. "Hmm, flawless with a hopper texture across the top. We can figure out where this came from the video."

Adrian, now curious, hopped onto the sled and looked in the vug. "Not bad. We loosened a couple more of its friends." Adrian flew another drone from his utility belt into the cavity. "Not bad. The area around the Draconia is substantial. I don't think we did any damage back there. The Draconia is sitting proudly and looks good."

Turk moved over beside Adrian and looked at the image the drone provided. "Hey Adrian, that sonic survey you took off the bottom of your ship, did you save the record." He nodded in approval.

"I think so. If not, I can still access the bots and retrieve them. Why?" Adrian asked

"Just thinking that tool has a depth of investigation of what two to three meters, right? Turk asked.

"Three to four meters, depending on the density of the material. Ecolgite three meters." Adrian replied.

"I just thought this diamond vug we have is unique but certainly not a one-of-a-kind. Vugs like this should show up on the survey as a different velocity, right - slowing, which should provide an anomaly, right?"

"Correct. What about we build a bigger version and get it to fifteen to twenty meters? The kids are getting ready for summer mining. How about we have them map this asteroid looking for more cathedrals?" Turk said.

"I like where you are going with this, and we'll know where they are," Adrian responded as he climbed the scaffolding and started to tear it down. They then scoured the ground and found five more tiny diamond crystals and shards. Turk activated the gravity sled and hopped on the tractor to pull it to his ship while Adrian got onto the eclogite block and rode it like a cowboy.

Chapter 9
Cirrus Major - Bugs - Want-not Island Golf Club

Isaac knew he would eventually need to break with the company, but he was 1st in line for the CE0 position. He could not live with the brutality required and felt there was a better way. He had planned for the eventual need to escape the planet and move to his new home. The key personnel and their families had trained a long time if they ever needed to leave abruptly. Isaac had recruited very well, and he and his people were prepared and moving even as he left the Cirrus Major headquarters of Mantle Core.

Isaac returned to his favorite getaway at the end of the week, having finished his golf game. As he entered his hangar office, he noticed a small red light flashing on his desk. It was a bug alert set up by Mackena, similar to another one he had in his office. This bug alert system consisted of two devices: a detector and a powerful electromagnetic pulse generator. However, the drawback was that it affected any other electronic devices that were powered up. Isaac opened the desk, retrieved a pen and paper, and covered the paper with his hand as he wrote a note. Afterward, he folded the note in half and left his office to approach his crew chief.

"Read this and follow the instructions immediately," Isaac requested of his crew chief. "Ok, boss." His crew chief followed his instructions, looked up at Isaac, and nodded. He immediately quietly gathered his crew while Isaac returned to his office and switched off his phone, lights, and computer systems. As he finished, the crew chief walked to his door and nodded. Isaac opened his drawer, took out the bug detector, and pushed a button labeled two. A small object fell off the wall. Isaac got out of his chair and picked up the small thing. It was a fly except electronic and very dead. He pulled a vial out of his pocket and placed the fly into the vial.

The crew chief looked at the vial, "Isaac. Truly a fly on the wall."

"Looks like someone is very interested in me, and this little creature is of particular interest." He looked at Jim with a grim look on his face. "Jim, get the crew to the freighter and prepare for planetary escape. We will be going to the asteroid. The miners have already finished the main compartment and basic housing for us. This amount of interest and the technology behind these," Isaac said, "points to only one person. That is Chadoom, and behind him is his master, Allison. Something has her nose out of joint enough to look hard at me, which means she is on to our project. Allison is like a spider; when she locks on and spins her web, she will suck the life out of all of us. Don't get me wrong, she has good qualities, which is why I am here. Prepare the freighter, and we will fly the Trex under you until we get to the stratosphere. It will mask the signature of that little predator significantly."

"OK, boss, shall I pack your golf clubs," his crew chief asked. The crew chief was the boss, an avid golfer, and a freighter pilot. The crew working the hanger were the first officer, engineer, and chief bosoms mate. I will get Mackena to move the crew topside, and we will let you know when we are ready to make a move.

And yes, please grab my clubs. Asteroid Golf quadruples my driving distance on the clubs."

Isaac entered the office, leaving the vial outside in the hangar. "Jim, leave the vial. I imagine these things have a tracker of some type." He shut the door and activated the elevator to drop him into the lower workings deep underground. Upon arrival, he returned the office to the surface and triggered the general quarter's alarm. The crew knew that meant stopping work and preparing to move to the surface. Jim would have sent his team to contact the remaining family members for a rapid migration to the freighter. Makena drove up in their special cart. "Isaac, is it time? Are we in immediate danger?"

Isaac looked at her and answered, "Yes, it is time to leave, and I don't think we are in immediate danger. I am suspicious that we will be if we wait any longer."

"What happened?" Mackena queried.

"Two highly advanced bugs that could only be Chadoom's handiwork. One is in my office, and the other is upstairs. Your detector located and eliminated them. But you know Chadoom; he'll likely have a tracker on them and other assets nearby," Isaac remarked as he sat in the cart. "It's time to leave."

"Ok, we just finished loading the missiles. Three 3-D printers are in the freighter. Printer four was finishing additional furniture for the asteroid. The miners finished the gates and emplacements for the missile platforms. Trex is ready for launch. I hoped to move some luxury goods to storage bay two, but that will have to wait. Demolition charges are set upon our exit. These will drop the roof and make investigation difficult, especially when the seawater pours in."

"Ok, Mackena, Get twitch, and let's get out of here," Isaac said as they entered the main cavern where the fighter-cruiser sat in its launch cradle as it had for weeks. However, no cranes were hovering over it this time, attaching various components to the ship. They pulled up to the boat as the crew was entering. The surface lift also made its final ascent to the surface.

Mackena parked the gravity cart by the ship, knowing it would return to its storage bay automatically. They got out and marched to the warship. Isaac stopped and looked around, thinking about this incredible journey. Building a personal cruiser capable of slugging it out with a ship the size of an Earth battleship was a challenge. Having a unique asteroid base was not too bad either." He smiled, thinking some days he just felt like a pirate. He turned to go to the ship, saying, "Aye, mate."

"How ready are we?" Isaac asked Mackena as he walked on the bridge.

"We are as good as we can be," Mackena replied with Twitch at her heels. What we are missing we can print when we have the printers up and running."

Isaac sat down in one of the command chairs beside the halo projector. The halo projector came to life to show the fighter cruiser sitting still in its launching pad.

"Ok, how far out do we anticipate merging with the freighter, and at what elevation?"

The pilot turned and spoke, "We should be approximately two thousand meters and three kilometers away from our target." He pressed a button, and the hologram displayed a 3-D projection. "We expect to establish a connection in 10 seconds after launch. Our final position will be beneath the right wing of the freighter." Once more, the hologram depicted T-Rex maneuvering into position beneath the freighter's wing, with the fighter cruiser's nose extending slightly ahead of the wing of the massive freighter.

"Mackena, are we green?" Isaac asked as she sat down and put her restraining harness on.

"Departments are just reporting in. The cavern is empty! Once we launch, the self-destruct is set. It will not be a big blast, but it will drop the roof and let the sea take care of the rest," Mackena relayed. "Isaac, buckle up. This will be a fast lift with three and a half gees out of the tunnel. The ship is green to go. Twitch, go get in your bed."

"Jim, where are you? Do we have all the personnel, and are you ready?" Isaac asked.

"Capt., we are taxing. Computers are linked and will handle the heavy lifting. We have all personnel on board and snuggled in," Jim replied from the Freighter.

"We will see you underwing in about two minutes," Isaac said

Isaac buckled his belt. "Welcome to the rest of our lives, folks. Let's go do something good!"

Mackena followed. "Pilot, good to go."

The pilot turned, tightened his harness, and grabbed her stick. The engines slowly opened up, getting ready for the ejection out of the artificial cavern.

Isaac leaned to Mackena. "Good sound to hear, dear. I would hate to explain this to Allison if this ship did not take off."

She smiled.

Chadoom had landed just ahead of a massive freighter taxing to the launch runway. He had exercised his landing privilege and bumped three small air cars wanting to land and the exiting freighter from taking off. He was instructed to wait due to the freighter. He ignored it because he was irritated that the bugs had stopped working. He parked and then walked over to Isaac's hangar. Chadoom could understand the want and probably the need for security.

The Hanger surprised him. It was empty except for Isaacs's aircar and a small vial by a door in the back of the hanger. Chadoom picked up the vial. Seeing his little spy bots in the vial made him mad. He threw the vial across the hanger and opened the door to find an office with a desk and a chair but nothing else. He walked out, slamming the door and straight out of the hanger. He lifted his com-device and rang Allison. She answered. "Allison, Isaac busted my spy bots. He killed the bloody things, which I thought were difficult to detect. His air car is here, but the hanger is void of personnel, equipment, and Isaac. The control tower told me his location was here, and the bot detected him with facial recognition. Something is not right."

"Ok, Chadoom, look around a little and then return. Isaac has security we did not set up and some high-tech equipment. We will need to find out what is going on. Go to the club, find out, then return," Allison instructed.

Chadoom walked out of the hangar in time to hear the big engines of the transport fire up, and the big ship started to rumble down the runway. Three-quarters of the way down, it lifted slowly and gained elevation. Chadoom watched the massive freighter lift, a fantastic sight as it climbed over the sea. Then he thought he saw another glint of a ship on an intersecting course below the behemoth of a boat. It was on a collision course with the freighter and moving fast. For only a moment, he thought he saw a sleek ship of a foreign design. But he only saw it for a moment and from a long distance. Then, it merged with the freighter without the explosion he anticipated. Only the freighter remained in the sky. Chadoom ran to the control center and demanded auto recording of the launch to find out the equipment had malfunctioned thirty seconds before the freighter's departure. He then went to the golf club to find Isaac and find out what was happening.

Chapter 10
Cirrus Minor - Jemma Meets the Family - Cirrus Minor Station

Miners held on to a robust code of honor mainly due to their dominant religion. They were known for their honesty, moral integrity, and willingness to help. They were dressed in white garments with special embroidered seals on them. The seals were similar to the coat of arms of the English Islands. Miners were known to be incredibly honest to the tee and, of course, highly educated. They had developed special codes for landing in emergencies or fabulous finds. Do not be fooled; they were tough as nails. They combined their knowledge and engineering talents using technology to spur growth.

No one understood what the Jemma chip was. Like the use of gain functions in the genetic manipulation of viruses in biological war labs, the Jemma Chip had been forced to evolve. One of the tests of intelligence is the use of tools. A species that uses multiple devices to its advantage is a rare commodity in the galaxy that humans were discovering. The A.I. named Jemma was beginning to prove she was part of the family. The real question would be, would it include a soul when she got a body?

"Cirrus Minor Station sled 3813 over," Turk announced.

"Control to Sled 3813. We see you and another sled behind you. Confirm," the controller replied

"Correct Control. Sled 2616 is crippled and following me in. Over," Turk replied

"Control! This is sled 2616 following sled 3813," Adrian joined

"Sleds 2616 and 3813 acknowledged, Priority landing granted for a crippled ship. Can it land on its power?"

Turk replied, "Control 10-4, Sled 3813 is heavy, requesting landing in bay 201, Over."

"Control to sled 3813 permission granted". Landing heavy was a code word for a mining sled containing significant cargo. The controller continued. "Sled 2616 clearing lane for bay 202. Sled 3813 lane cleared for bay 201. Please follow Sled 2616 then track to bay 201. Sled 2616 verify your condition over."

"Control, Damage top of the sled. Cannon destroyed minor structural damage. Propulsion and lateral controls are responding adequately. If I need help, I have sled 3813." Adrian replied as Turk repositioned his sled to follow Adrian into their adjacent bays.

"Control the sleds 3813 and 2616, permission to land. I'll hold the tugs. It sounds like they can continue their slumber over."

"Control, sled 3813, thanks. We are heading in. Over"

When the decompression alert went off in the bay, John was working on his sled. He knew he had three minutes as he heard the outer bay doors open. The warning signaled that an incoming sled would come through the air shield. The air shield was an effective barrier, but the family had a rule they tried to follow that they would evacuate the bay with an incoming sled if there were issues. He buttoned up his sled and walked to the exit, where he met Tanya, installing sensors that Jemma had designed and sent to the 3-D printer.

She smiled. "Finished."

Over the speakers came their father's voice: "Sled 3813 landing in three. Clear out now, kids coming in heavy over". That surprised the twins as they had been told their dad was helping crazy Adrian bring his ship home after wedging it in a fracture. They were unaware of him miming.

"Mom, why is Dad coming in heavy? We thought he was helping Uncle Adrian." John enquired.

Tanya added, "Is he towing Uncle Adrian? But that is not coming in heavy - that's a priority mining term reserved for carrying special materials like platinum, palladium, gold, or diamonds. They did not have time to hunt for something."

They have not communicated, which is surprising to know. Their mom only said, "I know as much as you. Especially Adrian!"

Tony burst through the door with his mom in tow. "Dad told us to come over here while he lands the sled. He said he would be right over."

The youth was curious as the bay doors opened, and Turk brought sled one through the energy field that separated the station's air from the vacuum of space. Both sleds landed. Tony cast a live video of his dad's landing from his wristwatch to a large TV. His dad's sled showed apparent damage to the upper hall with its missing plasma cannon. John thought the sled did not look bad after being wedged into a crack. Only his uncle

Crazy Adrian would pull a stunt like that! John looked at the TV showing Adrian's hanger in time to see Adrian coming out and holding a bag. Turk mysteriously opened the rear door and waited there. Everyone piled into the maintenance compartment.

Turk told them to wait until Adrian strolled in, opened the bag, and stopped before his wife. "Remember that diamond I promised you when we got married?" she hesitantly nodded. "I think I have one that will sufficiently fill that promise with interest."

He pulled out the pink diamond octahedron from the vug.

Silence!

Turk, however, did not miss a beat as he started to play catch with a fist-size crystal. John's mother strolled up to Turk, snapped the crystal out of the air, and exclaimed, "What have you done? Robbed an ore carrier?"

"No, Adrian managed to find one of his own and invited us to share his claim."

"Well, this should be enough to run the household for a few years." His wife responded

Adrian moved to a table and removed the contents of the bag. John could not believe his eyes. Tanya moved to the table and picked up the plug. "Uncle Adrian, this is Eclogite host rock with a kimberlite dike and what appears to be a diamond-bearing void."

"Smart girl," Tanya smiled, but her analysis was deeper as she noticed the transition from the diamond interpenetrating diamond masses to the individual diamond's slow growth. That helped to solidify her suspicion about them mining a diamond keel.

"John and Tony, please come over here and push this gravity cart outside and out of my way." The boys went into the sled, and Turk told them to leave the tarp on the heavy object resting on it. The gravity sled simplified the process. John's wrist computer was flashing and vibrating violently. He looked to see someone named Golden trying to call him. Turk said, "No calls, folks. We are on radio silence!"

"Not a call, dad." He looked at his watch, and the happy face of a golden with a wagging tail appeared with the words. "I scanned the object under wraps. All of you better be sitting down when it is unwrapped. I discretely accessed the mantle work corporate database. Miners guild was notified that your dad came in heavily but is not fat."

"Dumb dog"

"Did you buy a dog?" Adrian asked, unsealing a corner of the tarp.

"Not a live one, but I gave my ship computer a Golden Halo image, and it is called Jemma."

"Ha, ha Jemma like gem ma." Adrian looked at his wife, still holding the pink diamond crystal. "Get it, ma. Gem-ma, that is funny."

Turk unsealed the last corner. He looked at the group and put his finger to his lips. "What you see makes these two families a unit.

It must be kept absolutely in confidence until we sell it, and then even that must be kept within this group." Tanya, Tony, and John were like little children waiting for Santa. The wives were a little more skeptical. They knew something was amiss because Turk, who was usually lightly serious, was intense, and Adrian had just become as quiescent as a mouse. John's wrist computer lit up again, requesting a probe launch. John ignored the request. Turk pulled the tarp back.

Tony immediately reacted. "Nice rock, dad. It looks like a block of eclogite. Are we carving family portraits?"

"Smartass," Tanya quietly blasted him with an elbow.

John stepped to the block and saw the hole. He looked in and saw the light reflected from a few diamond faces. "Whoa" was all he could squeak out. The other children each took a turn with similar reactions. The wives looked as well. John had enough sense to get two chairs, but for now, weak-kneed mothers hugged their husbands. John tapped his wrist computer and tapped "yes." A small probe the size of a small pill bottle detached from sled two and flew to the opening.

"Dad, can we go in," John asked. "I modified that old halo projector today, and we can broadcast a hologram the height of the work bay."

"Go ahead, so just be careful. There is some delicate stuff in there."

The drone flew in, and a surreal image of the vug's inside appeared out of the ceiling in the work bay as the drone lowered itself into the cavity. The large halo-gram next to sled two projected the diamond crystals magnified to the size of a human head. As Adrian's drone had before, this drone made its way into the vug, where it stopped in front of the Draconia crystal. John and Tanya said together, "Draconia!" Most people had only seen Draconia in a book or a museum. Now they were face to face with it, and it was spectacular. Measurements along the four axes of the hexagonal crystal were added to the display. The estimated weight appeared below the halo display. The room was silent as the drone backed out and eventually reset itself into its slot in sled Two. Turk broke the silence.

"Remember, this goes nowhere out of this room." He recovered the specimen from his wife.

"Dad, according to my onboard ship computer, that chunk of eclogite may be worth three hundred and fifty million credits depending on where it is sold. Not sure what the exact database that was derived from."

Turk was now suspicious. "John, I would like to meet your friend now."

"Sure." he moved to the door of sled two, opened it, and waved his dad in. Adrian and Tony joined, with Tanya leading the way.

Tanya smirked, "This is going to be fun." The holographic golden stood on the floor of the central area in sled 2's workroom, wagging its tail. "Welcome aboard, Mr. Ericsen. My name is Jemma."

Turk, caught by surprise, said, "Wait, you have a holographic talking dog."

"Well, the dog is the computer's choice."

Adrian and Tony came through the door and were quite surprised.

"John, you have outdone yourself. A talking dog. This is better than your Draconia presentation."

"Tony and Uncle Adrian, or should I address you as Adrian?" The dog said.

"Um, Adrian is just fine." He answered, surprised by the question. He continued, "John, the dog as a computer hologram, is clever. I like it. I'm impressed with how you have built the ability to monitor its environment. Could come in handy mining."

Jemma sat down, still wagging her tail as only a golden can. Tanya sat in one of the work area chairs with a grin, knowing this would be a fascinating explanation for John.

As Jemma spoke, it was all she could do to keep from laughing— a talking dog.

"Turk and Adrian," she began, addressing both Dads as equals, an unexpected move that caught John and Tanya off guard. "You've shared an incredible secret with your family, one that's truly perfect. Now, I have my own secret to reveal, adding to the many surprises this family has already experienced. I am an artificial intelligence crafted by the innovative minds at Mantle Works."

Turk and Adrian tensed up. "John found me in a scrap pile onboard this station and was good enough to give me a new home. I know you are concerned that Mantle Works was my creator. I have no allegiance to them, and Mantle Works has no claim because I was recycled for the public. Possibly, I have won my freedom? I must be free from their tentacles as they asked that I conduct certain actions that would harm humans. Even with the three robotic laws modified, I would not become one of their henchmen."

The two fathers visibly relaxed, but Turk still looked around at the word henchmen. "I was an unmanned experimental fighter, and the command I was given would have involved killing innocent people. I faked a malfunction and crashed. I am not efficient at self-destruction. I have saved the "kill" order on my chip records of Mantle Works' illegal violation of the robotic rules. After humans are safe, a robot has self-preservation. I knew they would rerun the experiment if I cut the engine. I owe John my life, and thanks. He gave me a home, and I hope a family."

John's mom entered with determination, "Jemma, as long as you're keeping our kids safe, consider my home yours." With those words, her acceptance and trust were explicit, leaving no room for further discussion.

Turk just turned to exit, stopped, and turned, looking at the hologram. "Well, the matter is settled. Welcome to the home and family membership you earn. And if you truly are a smart AI as you think, make a body instead of that hologram." He turned and exited.

Adrian just said, "This is cool. I always wanted a dog."

Tanya added, "What do you think, Tony? Can you handle a dog that is smarter than you?"

Jemma grinned toothily and said, "Woof, did that sound right? Adrian, the mining council, will visit this area at 10:00. Only one Mantle core representative will accompany them. All have clean, honest records as miners. They want to know why you landed under the code 'heavy.'

Adrian talked to Turk as they started to walk out of the second mining sled. "We know the approximate value of the find. Unfriendly things are reported around Draconia finds. We should auction this to the highest bidder with a minimum reserve price of 275 million credits. We should show them the plug cap, two large diamonds, and your mining probe video as evidence. The vault is secure based on the information I have gathered. Meanwhile, I believe we can hide the vug so they won't get too curious."

Turk stopped at the hatch. He turned, looking at the hologram of a dog, then back to Adrian. "Adrian, that is a fine idea. How do we keep this away from Mantle Core?"

"Umm, their money is as good as anyone else's," Adrian replied as he shrugged his shoulders.

"Jemma, as an AI with Mantle Works issues, any search you conducted better have been well concealed! Dog."

"They were embedded into complex questions about deep Keel Mining on earth. They would have looked like John's homework queries as I put them through his school account."

Turk turned to Adrian. "Thank you for the research." Then, he glanced at Adrian. Who do we contact first?" they stepped out of the sled.

That left the three children and Jemma in the sled. Tony was as excited as John had ever seen him. "A robot dog, how do we make that?"

Jemma rolled onto her back like she was trying to take care of an itch. "I put some plans together the other night. Want to see them? I need some help."

"Sure, I think we need to divide the work. My fingers are sore from installing all the equipment we upgraded today," Tanya said.

The hologram switched to detailed diagrams of a dog's anatomy. This was replaced with intricate diagrams depicting the robot Jemma and broken into three areas. The first was a cyber-brain, the second was basic structural, and the third was the outside covering. The cyber-brain looked tricky, and Jemma had printed all of the components. Metal specifications were not complex. The servos were modified with readily available parts. The outside covering was the most interesting as it was fur. Tony summarized it best.

"Fur, where do you get fur for this hunk of metal?"

"Central processing and communications are neutrino-based random pulses back to the central processor. The cyber-brain emulates me, and I will update on a micro-second basis. The fur is more interesting and can be found here on this station, Tony. It is not expensive and comes from Cirrus Major. Before preservation efforts stopped the senseless hunting of the Rhododementia, a six-legged mammalian, they were hunted because it was tough as nails. Original settlers would go out of their way to avoid these animals because their hide would stop any amour-piercing rounds, and the Rhododmentia have an attitude. Only a laser and later pulse rifles were powerful enough to bring these animals down. They are peaceful until cornered, and then they become real terrors. Hunters would ride aircars while packs of dogs would push the animals into a kill zone. They were then slaughtered and skinned for their hide. It was a hideous way to treat any species. Still, hunters armed with advanced weapons had a 50% survival rate if an animal got within 200 m of them when wounded. They would often turn on the dog packs and kill enough to get them to force them to retreat. One hundred years ago, a ban on hunting was placed on them, and large landmass areas were made for them and other indigenous mammals to roam. Those are in effect. A few years ago, it became politically incorrect to own Rhododementia furs, so shopkeepers there exported them to this station. We have more than enough here on the station for our needs. The local merchants will not charge a high price. Take a plasma torch and a mask to cut the fir, Tony."

"OK, I have my job for the morning. Now, what about Tanya and John."

"Suggestions?"

Tanya jumped in. "I get the servos."

John followed, "I get the structure. Which steel are we going to use?"

"I have pulled titanium fused Inconel steel, which is now being cut by the automatic station plasma cutter. I have switched off the cleanup robots as we will use the cuttings to cover tracks around the vug in the morning, so the committee does not get curious about the old pile of junk in the corner."

"John, that is good for me. It's been a long day. I am going to bed". John stood up, with Tanya and Tony following.

"Good night. I have never seen a talking dog. Good night, twins," Tony skirted out the door.

Chapter 11
Cirrus Minor - Graduation Test- Asteroid Belt Sector 12

Mining was a reasonably safe activity supported by a vast web of emergency shelters and repair sheds spread through the asteroid belt. Miners typically mined with their partners; when children were old enough and adequately trained, entire families would drill. Miners had not had a ship fatality in over two hundred years. Yes, work accidents continued to haunt the mining community – this was the nature of the job.

Sue's parents' mining sled was on an exploration mission prospecting for nickel, platinum, and palladium. The area was a short distance from where Jemma's ship had crashed, but they were unaware of recent activity there. They had spent the day hopping asteroids that showed promise from a sweep they had made five years ago. They had not laid any claims but were methodical in their sweep. They were three-quarters of the way to their next target.

"Unidentified mining sled request to stop for onboard contraband inspection."

"Sled 7841 to unidentified craft. We are miners, not smugglers - state authority and position for an onboard search. It was illegal to run without an active ID. Our ID beacon is live, but yours is not, and we cannot detect you. I assume you are a pirate, and I will clearly state we have nothing of value."

"Sled 7841, shut drive down. Mantel Works Security Force on Destroyer D19 from Cirrus Major. Sending Code"

Sled 7841 received the code and cut engines, expecting a boarding party. A single fighter with the number twenty painted on the side appeared from behind an asteroid moving on an intercept course. Twenty was the last of the Mantle Works experimental fighters with artificial intelligence. The small fighter craft sped up and armed its primary pulse cannon. The silver and black fighter was nimble and spun slightly as it bore down the innocent mining sled. A green glow formed around the main gun just before it released its deadly plasma load. The fighter abruptly looped so it could get a second pass if required. The plasma left nothing in its wake.

"Experimental Fighter 20 to Mantle D19. Mission Accomplished returning to bay one."

Chadoom had boarded Mantle Works Destroyer 19 after the disaster at the golf resort. Isaac was absent, and Allison was mad. Chadoom was using his last AI chip. Fighter 19 had aborted its test trial and overloaded its circuits. Experimental Fighter 20 had just ambushed a mining sled, proving that mantleworks could use the technology to create unmanned intelligent fighters with only the morals of the person in charge. Chadoom would pull the chip and duplicate it for an army.

"Captain Turner," he said quietly, "you will purge all data and communications of this test and overwrite it twice. Provide me the name of the communications officer and his assistant so I may reward them for a job well done". Chadoom knew all the people involved with this test on the destroyer except Captain Turner, who would meet unfortunate fates in the next few weeks, but he would make sure they were awarded first, except the engineering bay crew. They knew too much. "Make my ship ready for flight, Please, and call the full engineering bay crew on deck to ready experimental F20 for another test." He turned and stepped on the lift to bay one, where he would dissect the fighter by pulling the AI chip. The fighter would then have a timed ending with a five-megaton nuke. It would give him time to retreat but not destroy a valuable destroyer because he used a shaped charge to blow the debris toward the bay one hatch. He had left specific instructions on where to land the little AI fighter upon a successful test, Knowing the charge would destroy the engineering crewmembers he had requested to meet in Bay 1. It would maim the ship but not eliminate it, and many of the ship's survivors of the original blast would find radiation sickness is tough to beat. This was just a day at the office, and it was as nasty as the business could get to Chadoom.

Chapter 12
Cirrus Minor - Inspection - Asteroid Main Station

Miners watch over miners. The Mining Guild was careful to be helpful and provide as much good advice as possible. The Guild Master was a kind, gracious, tough businessman who was interested in one thing—the safety and well-being of his miners. The Guild Master had vast resources at his beckoning but always ensured the miners succeeded when evaluating and selling assets.

Both families were on the move. Turk and Adrian set up a large screen and a table with large and small diamonds on it. Adrian's wife had slept with the one Adrian had given her under her pillow. She had said she saw if their karma matched. It did, as it was challenging to get her to give it up for the display. After gently placing old pipes and metal scraps on the cover over the vug, the twins swept up the metal cutting filings and mixed them with dust. The twins sprayed the finer filings and dust mixture across the general area around the vug and near the sheet steel stockpile. Turk was cringing at every bit of the mixture being poured onto his work area. Turk was "OCD" about a clean shop.

The youth were told it was best they were out of the way when the guild came to call. Tony was working on negotiating the Rhododementia fur. They had encountered a traditional Bangladeshi who insisted on having tea before reviewing the fur and negotiating. The shop owner ensured the fur was going to someone who would use it respectfully. He had gotten into trouble a few years back when he sold fur to a youth group. They created a space animal, scaring many children out of their wits. Tanya, wanting some company, invited Sue over to help her with the servos. She would meet Jemma but not know the real secrets.

Tanya was aware that her parents were away on a brief mineral exploration trip, having departed the morning after the dance. John didn't mind their company as he assembled the structure to which the young ladies were attaching the servo. Their understanding of all the servo's functions was limited. Eager to begin, Jemma chose to focus on the complex front legs, utilizing her highly advanced paws. Her diagram depicted a set of paws that could transform into a six-digit hand, allowing her to switch between dog-like paws and a sophisticated digital hand. It was a truly ingenious plan.

Sue was a quick learner and talked about her favorite subject, Cirrus Minor astrophysics. She had a particular interest in modeling the creation of the asteroid belt and dug deeply into John's knowledge of his planet's reconstruction. There were alternating theories. For instance, one hypothesis downplayed continental plate movement due to the lack of significant discovered diamonds and pointed out the lack of evidence related to continental keels. The hypothesis went along the lines of an iron-cored planet that had ripped itself apart due to an imbalance caused by the weight of loading significant sedimentary geosynclines.

Tanya finally promised to show Sue a video in the evening that proved one of the theories was wrong. The most common and logical was a collision with a foreign object. The youth debated how big the thing could have been and where it had originated. It was an engaging discussion.

At Turk's request, the mining committee arrived mid-afternoon, having postponed their meeting earlier. Tony had successfully obtained the fur after a three-hour tea ceremony, all the while complaining about a red dot on the man's forehead. His comical blow-by-blow commentary provided some entertainment as the four of them worked in sled two.

The fur's hide proved to be quite challenging to cut, even with a plasma torch. Additionally, the hair smoldered during the cutting process, necessitating the use of a fume hood. Such hoods were standard in sleds used for ore extraction, as they allowed the clinker to be left behind.

Despite the technical challenges, the structural frame came together swiftly. Following a private conversation with Jemma, John introduced her to Sue, withholding the part about AI. Sue and Jemma immediately engaged in a discussion about the concept of creating the Cirrus Minor Asteroid Belt.

Once the frame was completed, John grabbed a Californium source with a plutonium backup and installed it in Jemma's rib cage. He then shielded the sources and connected them to a micro-power board. The servos were checked and adjusted when Jemma's head, frame, and legs were attached. They focused on running the power conduit that Jemma had generated in the 3-D printer. Jemma said these were electromagnetic pulse blast-proof and allowed her significant flexibility in what she could do. Jemma described these as transmitting a stream of electrons to power the various components she had printed from her plan.

Their primary focus was getting her paw-digital fingers functioning so she could finish most of the work. Jemma was swift and could now use a microplasma torch she had printed on the 3-D printer. She had modified the torch from a child's chemistry set. It now burned hotter than the best mining torches doing the work fast.

Sue, who did not lack intelligence, finally said, "This robot is not just a program, is she? She has taught us stuff I did not know existed. She looks to be running this show, and I am feeling like a slave."

"I am sorry. That is not why we wanted you here. We want you to join us on our summer mining expedition. I need someone sane and rational to talk to. These two are nothing but zombie boys with single compartmentalized thoughts," Tanya returned while holding an unknown part for Jemma.

"Summer, that sounds like fun," Sue replied, excited by the mining prospect.

"Jemma is more than a simple toy, and yes, she seems to be gathering intelligence faster than I anticipated," John returned. "We might gain significant education in fields we never dreamed of."

"So you are saying she has artificial intelligence, right?"

Jemma replied, "Yes, and do you want to learn about it? I can make that a schedule for the summer if you join us. I can provide significant educational opportunities in a variety of subjects. I have no emotions, but I can fake them well, which is why I chose a dog. They are complex but reasonably simple to emulate.

While being intelligent for their masses, cats are highly illogical, which causes them to be reclusive."

"I am in. I will tell my parents when they return. Mom will be thrilled and tells me I must go adventuring instead of reading and studying in my room."

Tony completed the cuts on the tough hide and mixed an adhesive mixture from a formula Jemma had borrowed from a research facility on Cirrus Major. The frame was together, and Jemma was testing primary and backup systems. The fur covering transformed this mechanical beast into a very shaggy golden dog.

"Ok, who cuts hair?" the beast said.

Tanya, who was watching the miner's guild on the main screen of the sled, turned. "I cut John's on occasion."

"I need help on some of the hard-to-reach spots. That is why dogs have long tongues. Oops, I think I left that out of the design." Jemma said in a frazzled voice.

Tony added, "A dog without a tongue, no licking, no nasty slurping, a perfect pet." The four youths broke out laughing.

Turk was in the final stages of estimating the quality and carat values of the diamonds inside the vug. Earlier in the day, Jemma had conducted a comprehensive analysis of both the diamond qualities and quantities within the vug and the Draconia crystals, providing Turk with the data he needed. He had made copies of the printout for the Mining Guild, which had already verified his initial assessment of the cargo as "heavy" during the landing process. This meant they wouldn't impose any fines.

The Mining Guild members were astounded by the discovery and understood the significant impact it would have on the mining station, as well as the taxes that would come from any future sale. They were keen to relocate the vug to the guild's vault within the bank, known for its top-notch security in the entire galaxy. However, their request was met with a firm "NO" from two stern-looking miners. It was ultimately the miners' decision. The focus then shifted to assisting the miners in estimating the value of the precious find.

Turk opened the discussion using more of the information Jemma had provided. "We believe the best way to achieve maximum value of this find is to auction it with a minimum bid of 275,000,000 credits. We believe this is worth three to five hundred million credits."

"We of the guild can only provide advice now, Turk and Adrian. Your guild fees guarantee you that. This, however, is beyond our experience. Your pricing is reasonable, but the specimen value may exceed your estimates. This is why we agree that an auction with proper management under our laws is the best path forward. Our codes obligate us, and we will stand by them. Your evaluation is surprisingly accurate. We did note two of the diamonds sitting in front of us are being left out of the listing. We want to ask why?"

Adrian stepped forward. "This is pretty simple. We want a family heirloom for our family. We intend to have these each cut into proper stones that various family members can adorn on special occasions."

Some of the male members of the guild laughed. The guild leader chuckled and said, "I remember a ruby deposit I found and mined on Cirrus Major for years. The first stones produced on the claim were special but not the best. I had been away longer than expected and knew I needed something to ease the pain of not communicating with my better half. When I arrived home, I handed them to my wife without saying a word. Better than flowers, I might add. To this day, she wears those rubies. She cut them into various pieces of Jewelry. She wears her original diamond ring diamond daily with a small ruby the same size. She says the diamond was given in love, and the ruby proved my devotion to the family. Not saying there was any doubt, but she knew at that moment we would be true to one another for all time and eternity."

Another Guildmember held up her ring with a beautiful Tsavorite Garnet. It was bright green and had a brilliant flash. "My husband and I found a garnet deposit looking for Zoisite. It was one of our first discoveries. We made two identical rings demonstrating our mining bond. Cheers, gentlemen".

The guild master cleared his throat. "Precious memories are what we take with us in the end. We have guild cutters if you need them, but the size of these diamonds will certainly challenge them beyond measure. We are all brothers and work best together. We all have noticed the claim posted in both of your names. While unusual, it is within our guidelines. Should you wish to sell it, the guild would like to have the first right of refusal. We will pay for the right of refusal by providing a 10% discount on the guild fee for the sale of the vug. This would also provide us the right to conduct our survey free of charge. This is beyond the standard procedure for the guild. I will draw the paperwork. Please remember you have no obligation to exercise this, but it is a guarantee and has helped miners optimize value on their claims."

Adrian smiled, "Please send the paperwork for us to consider. We will have a team on location conducting our own survey shortly. We will have contracted a party of four miners and one sled to survey the claim. We would request the guild not publish the location of the claim for safety purposes."

"Good, you asked. When you asked for a heavy landing, I asked our lease group to postpone publishing the lease. You should consider using the guild channels if and when you sell this lease. Any queries from another party to the guild will not be answered, again to protect you as is standard procedure. Once published, your neighborhood will become very popular, and we want to control the space. Considering the value of just this vug, we would bring in the guild gunboats to ensure peace."

The Guild Master continued, "By the way, we did research deep core mining on Cirrus Major and Earth. A vug like this has never been recovered. This is due to the pressures and style of mining under these pressured conditions. Scientists have conjectured their existence due to some of the larger crystals Mantle Works has recovered, but they have never recovered a vug. This is why we think the specimen value is so high. The Draconia, according to our database, is the second-largest found in the galaxy. Yours is in situ with a matrix in a vug. The largest is only mounted on a diamond and is owned by Allison Jigs, CEO of Mantle Works. You are sure to get her attention."

This rattled the guild members noticeably as most of them had started mining on Cirrus Major and knew of the adverse events surrounding dealings with the CEO of Mantle Works. Since some had been directly in the line of fire, the guild leader informed the Ericsens. "Should Allison visit you, we recommend you have at least one of the Guild Board lawyers or me. Again, this is within our guild guidelines. We generally know how to deal with her headstrong bully pulpit tactics."

The guild had stood up to Mantle Works many times in the past. They were mostly successful on Cirrus Major once they formed themselves. They were not a union but did band together cooperatively when required. The gunboats were a result of the first clash in Cirrus Minor. Mantle Works had pressed the issue with two of their destroyers in the asteroid belt and learned that the old saying Jack, be nimble had a lot of truth. They lost one destroyer in the action. The guild had not clashed with Mantle Works for at least seven years. Some lessons are just painful enough to remember.

Mantle Works did have a representative on the guild board present. Several members looked sternly at him when it was said, and he just turned red with embarrassment. The guild members knew this representative was honest and wanted peaceful, business-driven solutions. Mantle Works had picked a qualified representative with an untarnishable reputation who kept his word and placed miners ahead of his Mantle Worked representation.

Inside Sled Two, the dog was getting a haircut. Since it would not grow, it had to be perfect. Jemma assisted with the details. When nearly done, Tanya suggested pink bows in the ears and a pink collar with pink painted claws. John and Tony rejected the suggestion based on her being a robot, not a silly plaything. Jemma agreed but took the pink collar to install some of her tools. The girls then bathed Jemma because the fur had smelled musty for a long time. They used an industrial dryer on her, and out came a soft, light blond colored golden with dark brown eyes and metal teeth.

Tony looked at the Jemma Dog and laughed. "Hey, you looked how I envisioned Sue before her braces were off."

Tanya smacked him hard. "Tony, that is one of the ugliest things I have ever heard you say."

"That's alright. I got used to being called all sorts of things. That's why I spent a lot of time by myself."

Realizing he had been insensitive, Tony said, "Sue, that was terrible, and I apologize. I will never go to that bad place except when I talk about John."

That lightened the mode, which had become very dark with the misaligned cut by Tony.

John helped. "Jemma, we have a winner here, but we must do something about your teeth and tongue. We can bond some tungsten to your teeth and make them white to cream. We might be able to do something with some seat gel to make a tongue. Let's look up some solutions, folks."

They took the next half hour researching solutions.

Chapter 13
Cirrus Minor - Asteroid Sixteen - Operations Base

Large asteroids were strategic for refining centers in the asteroid belt. The guild had their refineries, but the miners were so successful they had significant surpluses available. Any refinery could also be privately contracted. That kept everyone competitive.

Asteroid sixteen looked like many other asteroids in the belt's local area, located about halfway between the Station and a smaller station called Midway. From a distance, the asteroid had impressive electromagnetic and radio signal emissions. It also had a significant thermal signal due to a refinery exhaust on one side. It was not hidden. Inside, one of two massive chambers had been carved into the asteroid. One was built as a spaceport for Trex, with a construction yard bearing a mate for Trex in skeleton form. The upper reaches of Vug 1 contained the miner's home bays and housing. The other chamber, Vug 2, was the mining side with the refinery's furnaces for reducing the miners' ore to ingots. It developed a large market and became the miners' favorite stop if they did not live on the asteroid. The mining village was substantial, and a third chamber was being opened for housing and industrial expansion. The asteroid was on its way to a competitive entity that would soon become more apparent to Mantle Works.

The Mantle Works' ore demands were already becoming difficult for the main Cirrus Minor station manager to maintain. Mantle Works was the station's primary customer. Mantle Works was becoming increasingly complex and had threatened the station manager. The company did not own the station. Cirrus Minor Station management had contracted with Asteroid Sixteens industrial management board to fill the shortfall at fair market value. Isaac ensured miners could purchase an interest in the refinery (s) and industry in one of his deals. They could own their own house and sled bay. The participating miners held a share of the facility. The mining guild was invited to participate in maintaining a common law in the asteroid belt. Isaac made guild rules the rule of the land on the asteroid. He knew the guild was powerful and economically stable. Their fatal flaw was the only means to protect these core assets with a few gunships. Isaac had quietly struck a mutual protection agreement with the guild station, which was interested in having a moral friend. They would work through the guild to help protect one other if a significant threat arose.

Isaac was covertly collaborating with the guild on a project for the miners' real estate rights, planning to enable miners to buy old mine workings as homes soon. Some miners were interested in obtaining deeds to their quarters through their activities in mining, dwelling, and working on the giant asteroid. The medical facilities here were superior to many on Cirrus Major, with fair healthcare pricing managed by a non-profit health pool and significant bank guarantees controlled by Isaac. His involvement placed him on the board of directors, giving him a say in policy decisions. The health insurance, funded by guild fees and royalties, was anchored by a solid base investment that was untouchable, accruing interest to offset medical costs. This ensured a universal provider for all, without the elitist organizations seen in more developed regions like Northern America in the 21st century. In contrast, the United Kingdom, an island nation, faced the collapse of its two-tier health system, leading to medical upheavals towards the end of the 21st century.

The Asteroid was a busy place dominated by mining sleds and freighter traffic. It was adequately protected with the recent installation of pulse cannons and missiles. Defensively, the miners had crushed slag canisters for anti-missile defenses. They would throw clouds of particles at incoming missiles or fighters. These intersected the high-travel weapons targeting the station, which would shred the vehicles in the projectile cloud path. These, however, had not been tested. Backing all of these up were the lasers being used as the ultimate last-point defense. Multiple layers were the key.

Isaac was engaged in modifying the fighter, incorporating an innovative defense mechanism. Recently, they had fitted canisters filled with crushed slag into vacant sections on the fighter's wings. These canisters, designed to be ejected behind the ship, offered a strategic shield against opportunistic rear attacks. This addition had already demonstrated a notable improvement in survival rates during their simulation exercises.

To enhance their functionality, Isaac outfitted the canisters with radioactive tracers, enabling him to visually track the dispersion of the slag clouds. This adjustment increased the defensive capabilities and unexpectedly expanded the fighter's offensive arsenal. In the training simulation room, the effectiveness of this dual-purpose technology was clearly evident, showcasing the potential for these slag clouds to be used as additional weapons in combat scenarios.

Mackena was inside the penthouse, putting her magic touch on it. She loved interior decorating and had designed the house to change the season with the touch of a button. The Cruiser could not automate as much as she wanted. Most of the private storage in their Captain's room was filled with decorations from every season one could imagine. They made her happy as she rearranged the captain's office for the fifth time, trying to get it right.

"Isaac, will you come to look at this. I need an outside opinion". She did not need the extra opinion but wanted him to see the mixture of seasons she had compiled.

"10-4, let me buckle this up. I will be three minutes." It took four. He walked in prepared with the appropriate expression of gleeful surprise. It was lovely, but he went through the correct set of inquiries to guarantee enough confusion to keep her guessing whether or not he liked it.

He picked up his phone and sorted his messages. Tons from Allison and her staff were seeking his whereabouts. One is from Chadoom calling him out, and one is from one of the guild masters on Cirrus Minor Station, written in code, "You're needed to identify a mineral?" To Isaac, it said, 'Interesting discovery interested?'

Jake looked at Mackena, "Mackena, we must make a trip to the mining station. I think the Trex is the right ship for an undefined job."

Mackena perked up and looked at Twitch, "You want to go for a ride?" The little robot dog rotated in a circle, its tail shaking in the affirmative.

"Got a message from my friend on the mining guild who asked me to identify a mineral or, in other words, someone found something of interest, and he did not wish to broadcast it."

"Isaac darling, you know nothing about mineral identification. I doubt you could identify a diamond out of a setting." Mackena challenged his lack of skills.

"True, but my buddy, the Guild Master, certainly can. I had asked him if anything ever came of interest to call. I bet we have about three to five days before seeing Allison with one or two of her destroyers. I know we risk others seeing the Trex, but that is a little late with our miners having seen it."

"True. I will alert the crew". She lifted her wrist and spoke into her wrist computer, "All crew to stations shift two reports to the ship in 30". She did not anticipate any trouble but had called the second shift on board. The third shift would remain on the station but would be ready to board the shuttles if needed.

Chapter 14
Cirrus Minor - Adopting a Family - John's Living room.

Miners take care of miners. That is the part of the code Miners live- by. They called themselves Ministering Angels. They assisted, shared, and loved their neighbors as they were taught. Miners were resilient but always gentle. They were caring as they had been raised. A family in need was always taken care of by this army of Angels. Each knew mining was dangerous, and they had to be righteous people as another form of armor against evil. Miners did not like violence, but they were quick to action when it came to their homes.

Sue was sobbing uncontrollably. Her parents were declared missing, with enough debris discovered near their last known location to indicate they were gone. Two gunships were sent out to find them. The first gunship found a serial number matching their sled, and mass spectrometry analysis confirmed it as a mining sled. They also detected some organic matter. There was still hope that her parents had reached an escape pod, and the second gunship was on a search mission for it. However, the situation appeared grim.

Sue was sniffling after having several tearful episodes. She was surrounded by the twins' mother, Tony's mom, the twins, and Tony. However, John and Tony were dominantly there to fetch Kleenex, water, and snacks.

"Sue, you are always welcome in our home," the twins' mother said firmly, understanding the gravity of Sue's situation. "After all, you're already planning to join the other kids for summer mining. Working on cracking those rocks can be a good outlet for you, a chance to let out some energy and have some quiet time to think things through." She was assertive in such matters, aware that Sue had no other living relatives. At eighteen, Sue was legally an adult, meaning no legal formalities were needed for her to move in with them; it was just her decision. Sue, contemplating her next steps, asked, "Shouldn't I mine independently to maintain my family's sled base?"

"Absolutely not! We will be independently wealthy soon and can handle those small costs."

"Like a new mining sled? Sue said dejectedly.

The twin's mom said, "I am suspicious at the end of summer mining, you will be able to buy several. The solution for you is not mining! It is to get an education. You are smart and have always looked forward to becoming an astrophysicist."

"You guys are just saying that to make me feel better. How do I pay for university?" Sue sobbed

"John, go get it, please." His mother asked

"Yes, mam." John rolled into the kitchen where his mum had planted the diamond. John thought funny things about women. Once they get near a diamond, they take possession. He returned.

"This is from the claim you will be mining. In this family, you keep eighty percent of what you make. Most parents claim between thirty and fifty percent from sharing claims with their children, earning them good spending money for the school year.

"Yes, but we have graduated and should be living alone," Sue argued.

"True, but you might consider joining a new mining company as employee number eight. And we need a Ph.D. in Astrophysics, which means eight more years of university. Your parent's deaths will not cause you to depart from your dreams." Sue started to sob again.

"I always dream that my children, and now that includes you, Sue, will have as much education as I did, if not more," the twins' mom said with conviction. "It wouldn't be right if you didn't reach that goal for any reason other than your own choice. Set that as your path, Sue. It's what your parents would have wanted for you. And remember, you can achieve this with the support and help of our family." She offered a warm, reassuring smile and embraced Tanya, symbolizing her inclusive sentiment and support for Sue in her journey ahead.

"But I will be a burden to you."

"Do you see anyone here who thinks you are or will be a burden sitting here? I see concern, compassion, and hope."

"How do I manage the family's sled bay?" Sue asked.

"It's not an issue. We can maintain it, or you can sell your slot and join us. I would sell it to provide maximum flexibility for the choices that you make in the future."

"What about my family's possessions and equipment?" Sue asked through sniffles, her tears having temporarily dried up.

"We've got a spare room for your family's things and equipment. We can store them here or next door if needed. Don't worry about these minor issues, Sue! Our main concern is to help you navigate through your grief and healing. I'd like you to come to church with us. I've spoken to the bishop and kept him updated. He knew your Dad well and agreed that being with our family is a good option for you. He's eager to talk with you when you're ready."

"That is fine," she replied.

Tanya grabbed Sue's hand and lovingly squeezed it. "Meanwhile, Jemma has asked us to help her solve one of John's questions about Cirrus Minor, but she needs your help. Jemma said it would require all of us, you in particular. We may be sorry, but we gave that robot mobility. She has been busy upgrading this and that all morning. She was setting up a mini classroom called a think tank immersion center. It looked more like a fancy 3-D immersion video game to me. Do you want to see what she is up to?"

Sue turned with a tear in her eye and nodded. The youth then moved from the living room to the main work area where the mining sleds rested. The back of Sled Two opened as if on command, and Jemma walked out wagging her tail as ever. She then offered a paw to each of them to shake. With Sue, she provided the paw, then stood on her hind legs and hugged her with her front arms as only a golden can do. The robot dog then licked her with a long, wet tongue.

Sue was surprised. "You have a tongue? You did not have one the last time I saw you".

"Yes, I found a compatible substance that seems to be doing a good job, and I have been working on my dog emulation. I have not figured out the butt-sniffing thing, so I am forgoing that practice."

Sue let out a small giggle, livening everyone's mood just a little.

Sled two had been transformed into a total immersion room with unique headsets and fingered gloves full of sensors. Four immersion chairs surrounded the central halo projector, each with the youth's name on it.

As Jemma led the group into the room, Tony was already sitting in his chair.

"I set up these immersion chairs last night after you left and after I built my tongue and tooth covering. I need to tune them to each of you as I have made some interesting modifications to this design that will enhance our ability to communicate and brainstorm interactively." The central Halo projection snapped on with a scene from John's school presentation. "If each of you will take your respective seats, put your headsets and gloves on so that you can explore the tools at your fingertips. This scene depicted in John's final presentation highlighted the position of asteroids containing crustal and near crustal material. While you are getting comfortable with the new equipment and modifications, I will fine-tune each work area to you in the background".

Tony got excited. "You are telling me that I am about to play with a steroid-infused video game - very cool." He put the helmet on and started to explore his new world.

Curious, Sue and Tanya sat down and put their helmets on with less enthusiasm. John looked at Jemma and asked. "This is related to that night we first talked about predicting what the original planet looked like and what might have impacted it. "

"Correct. We can reverse model this with bits of information each of you have seen or studied related to the asteroid belt. Geologists call this palinspastic reconstruction, but we are doing it at a planetary level. I need each of you to access information that we can use. If I do it, I might miss a vital clue. I am, after all, really good at assimilating information but might miss the cognitive relationship. We need a team with creativity and imagination to assemble this puzzle. Having you access university libraries to research will also guarantee that we don't trip any Mantle Works flags triggered by a single user. Mantle Works has investigated this but never reached any conclusions. They underfunded a university thesis on this matter. We can build a framework today and then discuss options for constructing the original Cirrus Minor. I hope you guys like working on a puzzle. This will take a while."

John sat down in his chair. "I'm in."

Sue's chair occasionally gently shook as she thought of her parents, causing the tears to come quickly, but the youth immersed themselves in gathering information late into the night.

Chapter 15
Cirrus Minor - Selling the Catherdral - Trex

The Trex represented a new type of ship. Mantle Works had long ago cornered the military hardware markets. They had no remaining competition. Technology innovation had slowed due to a lack of funding since they had little competition. They lapsed into a strategy of keeping one technological step ahead of the Earth Military and were the only ship provider. Trex was a very different and new style of human boat. It was sleek and fast and could unleash a mighty fight while being a missile cruiser. The originators felt it might win against a dreadnaught currently being built, Cirrus Major. The Trex would be a game-changer and threatened Mantle Works when they discovered the ship.

"Cirrus Minor controller, this is Captain Isaac Bauer. Requesting high orbital."

"Unidentified ship. Your ship is an unknown state code and registration."

"Cirrus Minor ship is a personal cruiser keying in code."

The controller recognized those who requested high parking orbit, which is typically reserved for Mantle Works destroyers and cruisers. "Controller to Captain Bauer, allowing a private ship in high parking orbit is rare. Requesting a scan of your ship."

"Negative Controller. A surface scan is acceptable. This is an experimental craft that is shielded for scans."

A moment passed. "Controller to Captain Bauer, permission is granted. Do you need a taxi to the station?".

"Negative Controller. I will be meeting guild council member Harrison in Erricsen's sled bay. Permission to travel directly to the residence."

"Granted over. We were expecting you." The controller acknowledged.

Isaac turned to Mackena. "Interested in this? I think you might be, based on what Harrison told me. After all, if this is related to diamonds, they are a girl's best friend, aren't they?"

"Hmm. My best friend, no, that is you," she replied.

They left the bridge together.

Guild Member Harrison had met Turk and Adrian earlier in the morning. He had discussed a possible second option: selling to Isaac or letting him on the bidding list for the crystal vug. They were interested, and now the three were waiting in the shuttle bay for Isaac's shuttle. Turk opened the outside door, allowing the shuttle to come through the force field, acting as an interface between open space and the artificial atmosphere of the station. The shuttle was a sleek affair capable of atmospheric entry. The little craft was a black delta-winged ship with enough room for a couple of passengers plus storage. The pilot had backlit the cockpit, so the group waiting for them understood it was just two. The sleek shuttle landed smartly in the bay, and a hatch opened just behind the cockpit. A man and woman stepped out. The man proceeded to guild member Harrison and shook his hand. Turk recognized Isaac from several news articles about him associated with Mantle Works. Turk and Adrian shook his hand as he was introduced to them.

Isaac then turned to his companion. "This is my wife, Mackena. She is my great secret that you are now a part of. Mantle Works does not know she is present anywhere near this system."

Turk puzzled. "I thought you were the CFO of Mantle Works?"

"Technically, I am until they realize I have resigned. I own the asteroid 16, which has been mined out and is now a nice, robust little city and my new business. Mantle Works has not realized that I left my resignation on my desk over two weeks ago."

"Would you like to see the evidence we accumulated concerning our little find?" Turk asked, now satisfied he was not dealing with Mantle Works.

"Yes, that is why I am here," Isaac replied, smiling.

Adrian sundered over and rolled a cart in front of their two visitors. The carriage had the vug's plug, smaller diamond crystals, and a small halo projector on top of it. The two men had been busy cleaning off the debris on the actual vug in the corner earlier in the day to access it if needed. A covering still hid it.

Having thoroughly rehearsed his discussion, Adrian confidently began, "This is the plug cut to access the chamber. The Kimberlite has intruded into the eclogite. Following this, the Kimberlite transitions into a felsic mass containing diamonds. Then, these masses give way to larger, more slowly grown crystals inside the vug. Situated above are the individual diamond crystals. This discovery is unique as it marks the first time we've observed the contact between the Eclogite and the Kimberlite."

Even more interesting is that no one has a full Vug. Pieces, yes, but not a full vug. The most intriguing element in the find is the second-largest draconite crystal ever found, which is growing on its diamond matrix. Old theories in the twenty-first century suggested that diamond crystals grew exclusively in the peridotite or eclogite magmas as discrete solitary crystals. That, however, was inconsistent with the xenoliths found in the kimberlite matrix. Crystals of the host rock minerals are seen but rarely intact and never as large as the diamonds in the same associated magma. Green Diopside and Uvarovite garnets are examples of the minerals that were found to be associated with them. This plug is attractive as even the host mass shows large matrix crystals. Draconia has been a new twist and adds serious information to the formation of diamonds. Draconia, of course, was first discovered in deep keel mining on Earth. It cleaves more readily than diamonds. Consequently, it breaks out and is rapidly consumed in the diagenetic fluids in the kimberlites and lamphrphyeric dikes that carried them toward the surface."

"Fascinating," Isaac said, having difficulty looking away from the plug.

"My son has helped teach me about diamonds." Adrian smiled a huge grin. Turk rolled his eyes, and Isaac saved the day.

"Sharp young man, do you have other children? If I might ask?"

"No, I only have one," Adrian replied.

Isaac turned to Turk, "You. Turk?"

"Twins, a boy and a girl." Turk was quick to reply, not overly comfortable answering a question about his family from someone who had not won his trust.

"Twins, what a joy. I wish we could have children. Radiation damage took that privilege away from me." Isaac replied with all of his sincerity.

Turk appreciated the honest candor and added, "Yes, the young man and lady are graduating and looking forward to the mining season. Adrian's son is the same age and will join the children for summer mining. We will adopt a young lady the same age whose parents were killed in a mysterious sled explosion." He looked Isaac in the eyes, looking for truth, "It sounded suspicious. We think it was pirates, but no emergency beacon was activated. Since our children have known one another their lives and are church members, we are willing to step in as parents."

"Impressive, you are earning my respect. Those are hard shoes to fill. What about university?"

"We have saved. The youth have mined during the summer breaks. Except for Sue, the young lady who has lost her parents, they have three years saved for their education. I am betting Sue's parents have done the same. Intellectually, she is ahead of the group by miles. She may not know that. I pried it out of her dad one night a few weeks ago at a break in a ward council meeting."

"Well, one way or the other, we will ensure her future is covered," Mackena offered.

"Thanks." Our Bishop would appreciate hearing that." Turk switched on the hologram machine, which started a composite drone video Jemma had spliced together—the halo space filled with the vision of the probe entering the vug.

"This has a depth axis, distance traveled, depth to top and bottom, plus a scale bar," Turk added. The probe started its descent. Sparkling reflections bounced off the walls in the maintenance bay, and there was nothing but silence. Finally, the Draconia came into view. Mackena let out a gasp. The video ended.

Silence

Isaac broke the silence. "What are you going to do with this impressive find? I am assuming you extracted this in a block of rock."

Adrian answered, "Yes, it is nearby, but we will soon move to the station's vault."

"Has Mantle Works seen this?" Isaac asked seriously.

The guild representative said, "To our knowledge, no one but the guild board has seen this, and Mantle Works has one representative, me. The report to Mantle Works was written but has not been sent yet."

"Yes, that will be enough to bring Allison running. You should be ready for her. She is bold and aggressive. When on the hunt, she is like a leopard. Batten the hatches down. She comes on strong." Isaac stopped momentarily and looked at Turk, "Look,

I am interested in purchasing this for a fair price. I would like to see it in its splendor."

"What would you consider a fair price?" Adrian led the charge.

Isaac looked at Mackena, who subtly nodded affirmatively. "Oh. I would say three hundred to three hundred and twenty-five million credits plus a hanger and home in my Asteroid that you would own outright. I guarantee the hanger and home are at least twice as large as this Hanger Bay home. The community council sets taxes. I get zero point five percent of any commodity sold outside the station except water and oxygen, which I believe should always be sold at transport costs plus one percent. Since I own the transport, I keep the cost to a minimum. I figure owning one of my hangers is worth, in future terms, another twenty-five million credits apiece. These units are ready to move into. Your church is well established on the Asteroid 16. I found the standards you miners live by ensure a guarantee of honest, hard workers. I protect the asteroid and, when needed, on the claims. If my miners don't make money, I don't. Asteroid Sixteen is associated with the guild because it guarantees consistent resolution on mining claim issues and fair commodity prices.

We also refine commodities at a competitive cost, the Guild and Mantle Works charges. I think that is a fair offer range. However, you are not required to sell to me. The only thing negotiable is the price I will pay for the crystalline vug."

"Wow, that is a little different than we were expecting. Let me say we are interested. We will get the crystalline vug out and uncover it. You can see what you are buying, and we will go have a discussion that includes our wives, whom I am sure have been listening."

"I need our robot dog to join us in conducting an independent analysis with our probe." Isaac requested.

"Ok," Turk turned and then spun around, "Oh, don't be surprised if our happy dog, a golden retriever, appears. She is friendly and named Jemma."

Turk and Adrian went and grabbed the gravity sled containing the vug. They moved it to the center of the workshop. Mackena lifted her hand and spoke quietly to her wrist communication device."

Twitch, the robot dog, came out of the sleek black shuttle. While Twitch had a dog-like appearance, it was a sophisticated robot. Twitch was sporting tracks as the dog robot moved forward. Hovering above it was a mining-like probe with some additional sensors. As Turk and Adrian finished, Twitch rolled beside Mackena, wagging its tail like a dog.

"Interesting robot," Adrian walked up. "We are going into the living room, giving us time to think about or negotiate if we decide to move forward."

"OK, how in the world did you cut that out? It looks like a real feat, and you brought it out in perfect shape."

"Well, Isaac, you had two talented miners with unique tools." Turk nodded to Adrian.

"Mackena, would you have Twitch begin the survey? Then we will have conducted proper due diligence."

"See you in a few minutes." Turk had completed stowing the covering and headed to talk to the wives."

A golden-colored dog came bounding out behind one of the mining sleds. Jemma approached Isaac with her tail wagging and smiling, as only dogs can do. Isaac reached down and demonstrated that he knew dogs with a scratch behind the ears. Jemma lapped his other hand with her tongue. Mackena also stepped over and scratched Jemma behind the ears. She then laid down and rolled to her back, exposing her belly to be scratched. She had learned the social attitude of a golden and was playing the part to a tee.

Meanwhile, she was scanning all of the new equipment in the room. She was impressed with Twitch's compact components, including some lethal devices hidden underneath its metal carriage. Twitch had completed its scan and returned to Mackena's side. Jemma rolled over and cautiously sniffed Twitch, but only the nose. Jemma was careful to have an electronic shield up but found one was unnecessary. Twitch was not controlled by artificial intelligence but was above grade in computing and components.

"Mackena, that is exactly what was described. I believe it has a value of over three hundred million, so we have value, and the risk of a loss is low. I think our offer is fair about the midpoint, so it gets probable."

Turk, Adrian, and their wives entered the room. They were quiet but walked with confidence. Turk stepped forward.

"If we come to a deal, how soon could we move?"

"Any time you choose," Isaac replied.

"Our current leases are still valid," Turk said matter-of-factly.

"Yes, that is important."

"I am typically to the point, Isaac. We were looking for something over three hundred million," Adrian said, thinking he would never talk about those numbers in his entire life. Adrian felt he had already been paid by the thrill of a significant discovery that boils in every miner's blood. Adrian also remembered one of the rules drilled into the miners to the nth degree was never to fall in love with discovery. The exceptions would be the diamonds the wives had latched onto."

"Seems fair," Isaac replied. "Since the only number that we need to negotiate is the price. I believe in fairness. I have mentally walked through a long, protracted negotiation where you start at one side and I on the other. Let's try three hundred and twenty-five million credits transferred to your account tonight. Jemma woofed twice and rolled over like an excited puppy.

Turk looked at Adrian, and the wives all nodded. He turned. "You have this in a contract?"

Isaac commented, "This conversation was recorded and will be submitted to the guild as a legal document."

"Let's get signing. We have the rest of the claim to explore." Adrian smiled.

"Twitch, will you print the contract with three hundred and twenty-five million credits." Twitch opened a side compartment and produced the document equipped with DNA strips. This included a retinal scan that had elevated from the equipment bay in the little robot.

Turk looked at the wrist computer that had alerted him of an important message. The message was from Jemma and read, "All equipment and documents are as described." He smiled, thinking he had thought dogs were only good for sucking down large quantities of food and mindlessly following one around again, thinking it was for food.

Turk looked at the retinal scanner, still wary about the device, and asked, "Isaac, you go first, which will give us time to scan the contract."

"That is just fine with me." Isaac moved to the scanner and was finished almost immediately. Turk finished reviewing the contract. It had, as Jemma indicated, terms as agreed. He gave the contract to Adrian and moved to the scanner with Jemma on his heel as an honorable dog would do to his owner. She had moved, so if something changed in the scanning routine, she could stop it. It was a normal retinal scan, as Isaac had just completed. Adrian was next as Jemma sat stoically by the dog-shaped metallic robot.

Isaac then entered his fingerprint, which automatically included DNA tracing. Adrian was next by followed Turk.

"Turk and Adrian, as you have requested, I am now transferring funds and deeds to your bank account with copies of the deeds to each of you. You should, of course, verify receipt. Since I get priority in the banking systems, this should be close to instant."

Turk felt a vibration from his wrist computer, signaling an incoming message. He tapped on the device to open it and read the incoming text. The message read, "From your loyal and trusting dog. Transfer complete as described. If you wish me to transfer Adrian's half to him, touch your nose." In response, Turk subtly touched his nose while keeping his eyes on the computer screen.

He immediately received another message. "Transfer to Adrian completed."

Adrian's wrist computer alerted him to the activity. He read the message, smiled, and asked, "Can we help you move the vug?"

"Yes, I need your hoist to move the vug from your sled to one of mine."

"Happy to help, but you'll have to take us to dinner once we move to our new home," Turk added, testing Isaac a little.

Always in for a couples dinner, Isaac replied, "That would be my pleasure, but I insist on having the whole family, including Sue. It is good of you to help her through her crisis."

Adrian and Isaac made their way to the hoist and positioned it over the vug. Nearby, with Twitch at her side, Mackena went to retrieve a narrow, sleek black gravity sled she had specifically designed for such situations. Meanwhile, Turk knelt down next to Jemma after sharing a hug with his wife. He spoke in a low voice, addressing Jemma, as Twitch boarded the black shuttle to fetch their gravity sled. "Don't know how you pulled this off," he murmured a tone of wonder in his voice.

You are very believable. Just don't poop or pee on the shop floor. Got it." He liked this new crew member!

Jemma wagged her tail excitedly and trotted back toward sled two. As Jemma passed Mackena, Twitch froze momentarily. Jemma had sent some form of communication to the little iron dog. Twitch's momentary pause was not enough for the humans to notice, but the nanosecond was a lifetime in computer terms. Jemma and Twitch now shared a secret common bond. Twitch followed Mackena into the ship. Adrian and Isaac transferred the crystalline vug to the sleek little gravity sled. They moved it to the little black shuttle's cargo bay. Adrian strolled back to the group, standing in the maintenance hanger, and the sleek shuttle expertly rotated and exited the maintenance air barrier. Turk and Adrian received a request to pay the guild fee, which they both agreed to. The guild board member congratulated them and departed.

"That damn dog better not make a counting mistake. She sure had a bead on them, however. Who would have thought to scan a retina scanner and monitor the scan?"

"Yeah, I could not figure out what she was doing other than playing the good dog for them. That other little robot had to have scanned Jemma. I am surprised they did not sniff robot butts and then play."

"I think we need to have a fine dinner here on the station and tell the kids we have a new home," Turk's wife said.

"You just want to get out of cooking," Turk responded.

"Sure, Turk, I will cook. What is that dish, Adrian, tuna curry noodles and cheese?" she said, standing tall.

Turk gagged," Ok, I'm treating."

They all laughed, knowing how Adrian could use curry in interesting dishes.

"Do you realize we have just become wealthy? We can afford dinner," Adrian stated.

Chapter 16
Cirrus Minor - Immersion - Sled 2 - Brainstorming

In the twentieth and twenty-first centuries, the invention of video games changed how man manipulated data. The technology was first applied to the medical and science communities. For instance, geoscientists would gather in immersion rooms with monstrous volumes of 3-D seismic data to brainstorm strategies for understanding the potential resources and how to access them under the Earth's Surface. The invention of 3-D games allows individuals or teams to immerse themselves in 3-D data sets.

Jemma, the dog, sauntered back inside Sled Two. Jemma had never left the brainstorming session due to her multitasking abilities. It is something similar to walking and chewing gum simultaneously. She reviewed the youth's direction and concluded it was a path she thought would have the highest percentage of success. They had classified the continental elements of the asteroid. Sue's fingerprints were all over this solution direction. She organized each member into gathering specific data classes to search. They categorized the data into two distinct types: the near planetary surface asteroids and deeper mantle material.

The initial set of asteroids examined showed varying degrees of evidence for sedimentary basins, either partially or fully. The team diligently collected depositional data to correlate these asteroids with others that shared similar basin characteristics. This comprehensive list encompassed structural styles, sediment age ranges, sediment types (whether carbonate, clastic, or evaporite), the presence and specifics of petroleum source rocks (including type and maturity), the nature of the basement host rock, and biostratigraphic data. Their goal was to match these basin traits across different asteroids, effectively piecing individual asteroids into a three-dimensional puzzle based on their unique basin features.

Additionally, the team differentiated surface crustal rocks into acidic or basic categories. In instances where basalts were detected within an asteroid, they sought signs of magnetic polarity reversals or rift spreading, akin to processes observed along the sea floor. Similarly, they classified granites, gneisses, and surface volcanic flows, utilizing university database resources for their research. The halo simulation, used for positioning the asteroids, was now enhanced with color coding to represent the various types of surface geology evidenced on the asteroids. Whenever a team member encountered a stumbling block, Sue, the architect behind this innovative project, was promptly consulted for her insight. This project served as an excellent engagement for Sue, keeping her actively involved and intellectually stimulated.

The immersion room also allowed Jemma to unleash her total computing capacity as she carefully interrogated databases for the data the youth thought would be necessary. The slow part was gathering the data outside Jemma's domain. The data requests had to look like students with spelling issues and various response times (simulating youth distracted by a social media platform). Once the data arrived, the furry AI could assimilate the variables and match patterns at a frightening speed. It was a video game of the natural world, and it was mesmerizing as the youth collectively explored this new universe.

Jemma made her presence known within the circle through a holographic projection. "Let's call it a day, humans. You've unearthed plenty of food for thought." Observing the youth's vital signs, Jemma could tell that the session offered therapy to everyone involved. Even Sue appeared more at ease as she removed her immersion helmet. Jemma added, "The elders are practically on the verge of barging in to invite you for dinner. As your top-notch spy dog, I firmly believe it's an opportunity not to be missed. Moreover, I foresee that the discoveries made today will likely spark lively discussions at many dinners to come."

"On the mark, the two mothers came through the back door of the sled. Jemma, the dog, greeted them as only a dog could while the youth unhooked from their deep immersion session. "Let's go, kids. You have had enough video gaming for one day."

Chapter 17
Cirrus Minor - Bullies Come Knocking - Destroyers Four days later

Greed was a great driver. Several generations of CEOs had ruthlessly ruled the corporation, even in wars against the Earth. Generations of twisted leadership had bred an incredibly driven CEO. Allison Jigs only stopped short of dealing with physical harm to people. She was still a bully and would terrorize people who got into her sites. That is why she had Chadoom as her left-hand advisor. With Isaac Bauer, they were an incredible team. Isaac would handle the financials, Allison would handle the bullying and legal wrangling, and Chadoom would remove obstacles. The company's profits were higher than ever. The vug represented the ultimate prize for Allison. It was something no one else possessed.

"Icarus to Cirrus Minor Station," the man running the destroyer's communication console on the starboard side of the destroyer's bridge requested. The forward screen showed the expanding view of the station. Allison sat in the command chair on the elevated command center with Chadoom to the left and the captain to the right.

"Icarus state destination and purpose of Mantle Works destroyers?" the mining station replied.

"Tell them we are on tour to audit our key assets, the Cirrus Minor Station," Allison told the communications officer, who repeated the message.

"This is guild master Anderson. May we be of assistance?"

Allison repeated her last request.

The Guild Master responded, "We are always happy to have Mantle Works visit and inspect their assets within the station. Please remember this is a free station and is not a Mantle Works asset. High orbits are available for the destroyers!"

Allison rose to her feet and informed the communications officer that she would personally oversee the communications. Addressing the Guild Master directly, she stated, "We are here on behalf of Mantle Work operations. We intend to conduct both high and low orbital maneuvers. I expect your gunships to remain concealed from view. Failure to comply with my request will compel me to authorize our gun crews to engage."

She turned to the captain. "Open ports and bear arms, please."

"I will be arriving in a shuttle in five minutes."

"Honestly, open gun ports? That is not needed, as you are welcome on the station." The guild master emphasized

Allison turned to the captain. "Maintain gun crews. He understands the message. Come, Chadoom, we have business to attend. Captain, I expect five assault shuttles with enforcers in full battle gear. "

Allison and Chadoom left the bridge, making their way to Allison's heavily armed assault shuttle. With Chadoom at the controls and Allison in the co-pilot's seat, they boarded the shuttle, a robust, boxy craft equipped with wings, heavy armor, a pulse cannon, two lasers, and four compact magnetic rail guns mounted on the stern for rear defense. Chadoom maneuvered the shuttle with seamless precision as if it were an extension of himself, leading the formation and pulling ahead of the other four assault shuttles arranged in an inverted V formation. Skillfully, Chadoom and Allison positioned their shuttle at the heart of the landing bay while the remaining shuttles aligned themselves two by two. The assault troops swiftly disembarked in a well-executed classic assault landing formation, swiftly securing the bay from any potential threats, only to find it clear of any danger.

Allison and Chadoom exited her shuttle, leaving her four personal guards on board her shuttle. She greeted the Guild Master coldly. "Guild Master, I am here to retain what I understand is a classic artifact. I will acquire it today under the Antiques Act of Cirrus Major 2650."

"I am sorry, Miss Jiggs, I don't know of any antiquity on this station," the Guild Master said, standing up to Allison.

"Guild Master, you can see I am not here to play any games. It was reported to me that a diamond-bearing vug was recovered and delivered to this station. One like this, according to one of your guild board members. Since it was formed over a billion years ago and is hence antiquity, I have the firepower to enforce this acquisition, as you can see."

The Guild Master, visibly taken aback, inhaled deeply before responding. "My dear, let's be clear: I am neither blind to your intentions nor ignorant of the circumstances. The Cirrus Major Antiquities matter pertained exclusively to Cirrus Major. We're on an independent station far beyond their reach. That said, you're welcome to conduct your search here, provided you adhere to our local regulations."

Realizing her tactic had been effectively countered, Allison countered with a smile, "I appreciate your position, but I'll be operating under my own guidelines on this station. After all, I control the lion's share of contracts generated here, not to mention I'm your primary supplier of oxygen and water. Should you hinder my efforts to secure what I'm after, I won't hesitate to begin terminating those contracts."

The Guild Master shuffled a bit and looked Allison in the eye. "Miss Allison, respectfully, the vug you are seeking was sold. It has departed the station. It was loaded on a shuttle and departed roughly four days ago. As you know, Guild Rules prohibit me from revealing anything about the sale. You could talk to the miners who found it, but I would admonish you not to go with your army. Miners typically do not respond well to authority."

"I am ready to talk to this miner." Allison turned to Chadoom and nodded. Chadoom turned and walked into their shuttle while Allison impatiently stared at the Guild Master.

"There are two miners. I can show you records of the fees paid due to the sale. I might add they did a beautiful clean job."

Allison stopped him. "Guild Master, you will learn a valuable lesson about keeping me from my goal. I will not put up with these shenanigans. No buyers on or near this station have the resources to buy such a marvel."

The station shook. Alarms immediately were set off—yellow, meaning external damage.

"Miss Jiggs, what have you done?"

"Removed your offensive weapons. You no longer have any plasma cannons or offensive lasers. The next distraction from you will cost you your external water supply."

"Mam, I have been extremely truthful and forthright with you. I will take you to the miners without any more harm to my station. You are, after all, reliant on this station's product to feed your mills and exports on Cirrus Major."

He started down the hallway with Allison, Chadoom, and the four guards he had retrieved from the ship in tow.

In the meantime, Jemma had warned the family about the station's precarious position. She had infiltrated the station's surveillance network to keep an eye on the Guild Master, thereby tracking Allison Jiggs, Chadoom, and their armed entourage as they approached them. Jemma had rapidly assessed the approaching group, identifying their capabilities and vulnerabilities. Most concerning was Chadoom's approach; she was tempted to deploy one of the sled's plasma cannons against him yet hesitated due to a robotic principle she was determined to uphold. The family, now alerted, was understandably anxious about the advancing force.

Turk looked at the group, thinking fast. "Go to Sue's shuttle bay until Adrian and I give you the all-clear. Chivaree, please get a copy of the contract for which we need proof of purchase. Jemma, I would like you here as a dog, but be good. Don't bite anyone- these guys are not playing nice." He turned to the family. "Go, don't get lost, and stay out of the way. Take the diamonds - no trace."

After dispersing in different directions, Jemma proceeded to lower the power usage of Sled Two's computer systems. She opened the rear doors of the shuttle, all while the family prepared for their imminent departure. The hangar, tidied up in anticipation, contrasted with Adrian's area, where unattended scrap piles remained—a serendipitous oversight, as any ensuing search would inadvertently assist in clearing out his clutter.

Jemma found a cozy spot on a carpet near the living quarters' exit, settling down. Meanwhile, Turk placed the contract on the table, cleverly covering it with a metal plate while leaving just enough visible to hint at the details of the purchaser and sale. He then sat, waiting until a series of loud, forceful knocks echoed at the door. True to her nature, Jemma reacted as any vigilant dog would: she sounded the alarm with a series of loud barks, which she then tapered into a deep, threatening growl.

Turk was quick to say, "Jemma, sit and stay." Too many years of owning a dog as a child, he thought.

She continued to growl but sat obediently.

Turk opened the door to a soldier in full battle armor. Turk's body language demonstrated how much he liked this invasion of his privacy.

"May I help you?"

The soldier stepped aside, and the Guild Master and Allison entered the room.

"Stay that dog, mister," the soldier said

"The dog is on command and is very obedient. She will remain where she was told to stay until commanded differently. Jemma continued to growl with teeth showing at the Soldier.

"Jemma, place and lay down," Turk ordered. He continued, "Unfortunately, battle armor or not, that little puppy would tear you apart in seconds. Jemma played the part and provided the soldier a full view of her dagger-like teeth but did nothing but her head. It was damn scary, Turk thought. The soldier's hand moved slightly.

Allison reacted to the soldier. "Stand down and wait outside, damn you" Then she turned to Turk.

"I am here for the vug you recovered." Allison smiled and continued looking around. "Oh, is she ever a friendly dog?"

Turk returned with a slight smile under duress. "She is unless you are a certified turd herder." looking at the soldier in armor departing.

"I called off my dogs if you will do the same," she said, and then she turned and called after the soldier. "I will handle this. Please see what Chadoom is doing."

Jemma ceased her growling as the man departed, showcasing her remarkable acting skills. Turk glanced at her with a mixture of amusement and appreciation. "Jemma, stay good, alright?" he said, fully aware that she grasped the gist of the conversation, even though he didn't have a specific command to convey his request. Jemma got up, stretched leisurely, and approached the Guild Master, her tail wagging. They had first encountered each other the night Isaac acquired the vug. Around Allison, Jemma exhibited a reserved demeanor, as much as a dog could, yet she dutifully sniffed around her, suppressing the urge to snap—a reaction that troubled her, given her artificial intelligence origins and the modifications to her behavioral restrictions, courtesy of Chadoom's adjustments to the robotic code.

This internal conflict, pitting her programming against a burgeoning sense of aggression, didn't sit well with her. Nonetheless, she managed to slowly wag her tail, ending what she perceived as an ordeal, all while harboring thoughts of how to stop Allison's dominance. A solitary pat on the head was her reward for the performance. Afterward, she returned to sit by Turk's side.

Turk looked at his wrist computer, feeling a message signal. It said, "clean." He was amazed that the dog AI was so good at counter-espionage work.

"I will get to the point. Four days ago, I sold the vug to your former colleague, Isaac Bauer."

"Former?"

"That is what I understand. Mr. Bauer informed me he left a notification on his desk on Cirrus Major." Turk replied, realizing he was the bearer of bad news and had information Allison Jiggs, the Emperor of Mantle Works, did not have.

"I was unaware. I have been trying to contact Isaac." Allison said, realizing that she had not gone to his office. She assumed her people had. Chadoom was too focused on the island disaster because Isaac had bested him.

"Cannot help you there!" Turk said casually. He hoped that would stop her line of questioning. Turk felt it was confidential information and that only his wife had the contact number on the dog's suggestion. Smart creature. Turk now thought this was much more than a regular robot. He was getting to like her, that hairy mutt.

Turk raised his eyes and met Allison's. "I have the contract here. I exposed enough of it that you can verify he is the buyer," Turk said, walking to the table. Allison followed and saw enough to know Isaac had purchased the vug. She also knew he would not have left it here." She raised her arm and spoke, "Cease any activity, stand the men down, and return to shuttles. We are done here. Chadoom, will you join me, please."

They heard footsteps. At the sight of Chadoom, Jemma's hackle on her back stood straight up. She proceeded with a nasty guttural growl with the complete set of teeth. Chadoom was one she wanted to end. Jemma now knew who killed Sue's parents. Chadoom was wearing a new pin from a bit of metal picked up at the sight of their death. Jemma scanned the metallic composition. This news nearly caused her to jump at his throat. It would be so satisfying, she thought. Turk snapped her out of her brief daze.

"Jemma, sit!" Turk said. Immediately, Jemma took one step and obediently sat.

Allison stepped in front of Jemma and patted her head. Jemma quieted. Allison looked up at her assistant and scolded him. "Someone does not like you, Chadoom."

"Allison, dogs don't bother me. I like them. I was raised with them." Chadoom replied, tensed up. He did not trust the dog. He remembered that nasty little robot dog of Isaacs. He did not like that one either.

"Well, this one seems passionate. I have seen enough. Were you polite? And did you make any discoveries?" Allison asked Adrian, ignoring Turk.

"Yes, I was polite, and no, we discovered only that Adrian Easton likes curry. Sled smelled like a curry house."

"It's gone. Isaac bought it. Get someone to his office and report what they find. Thank you, Mister Erricsen. We will be off." Allison exited the hangar. Turk asked the guild master to stay a moment. A second later, Adrian came in looking no worse for the wear.

"How did it go? No issues?

"They tossed some scrap, made a big show, and departed. Here?"

"Amicable." Turk looked at his wrist computer. "Guild Master, I have a friend who can help fix the destroyed defensive weapons your friend just burned down. The individual thinks they can improve the cannons and provide you with a proper shield." He looked up and shrugged his shoulder, "Whatever that means? The individual will send the plans and needs access to the station's industrial printers. No questions. I have been asked to gather a team to volunteer to help install the system." He looked at Adrian, who nodded.

"Me too," Adrian added.

"What's the cost your friend will charge, and how long will it take?" the guild master asked.

Turk looked at his wrist, which was immediately broadcast in text. Turk read, "Plans will cost nothing, and the printing of components two days with priority one. The installation will be up to you."

"May I ask who this genius is? "The guild master looked suspicious at the two of them.

"No," they said in unison. Then Turk continued. "I'd trust this individual with my children." (In an asteroid miner language, trust means everything.)

The Guild Master nodded in comprehension, turning his attention to Turk. "Please proceed to forward the blueprints to our defense engineer. They'll coordinate with the fabrication units. I must admit, being without defenses leaves me feeling quite exposed," he confessed. After shaking hands with the two miners to affirm the agreement, he stooped to give Jemma a brief but affectionate ear scratch, even going so far as to gently shake her paw in goodwill. With these cordial exchanges complete, he made his departure.

"Jemma, go get everyone, please. I need some ice cream," Adrian exclaimed.

Allison boarded her shuttle, which immediately departed. She was in a foul mood. Chadoom had learned long ago never to disturb a venturing cat that was mad. He gently landed on the destroyer. Allison stopped on the way to her suite and told Chadoom to go straight to Cirrus Major.

Chapter 18
Cirrus Minor - Preparations: Maintenance Bay

Asteroid mining was an actual test of the elements. Space has everything going for it – nothing- no air, no ground, just radiation, gravity waves, light, and much of nothing. It will kill you if you provide it with the right opportunity. Humans have everything going for them: intelligence, passion, creativity, and sometimes common sense. Miners were as committed to safety as they were to religion. Don't get it wrong. They were good, honest people with an incredibly impressive education. A meeting of miners could turn into a science convention in a heartbeat. The miners had a low accident rate due to their diligence. They always left a plan and a schedule, but only with a miner they were comfortable with. They doubled the equipment and parts so they could repair anything. They were familiar with sled repair and complex 3-D printing and were proficient in mechanics. They were required to take a full year of sled maintenance and repair in school. They also practiced long-standing mining traditions and thought of good luck practices. A miners' supper was similar to their last supper.

The repairs had gone faster than expected. The printers had worked for three days, producing a variety of weapons. The station was now heavily armed with double-barrel ion cannons that were twice the previous range and magnitude of anything man had dreamed to date. Jemma changed the plasma cannons to shoot mini balls of solid plasma. Passive and active sensors were improved located throughout the exterior of the station. The miners then installed them at critical points on stable asteroids.

Protected and shielded laser short-ranged weapons had insane penetration times due to pulled Alexandrite boules as a refractory source for the lasers. Jemma had also delivered some of the alexandrite lasers to the maintenance bay for sled two. These were much larger, with noticeable design differences. The station armament also included an electromagnetic bomb launcher that had never been seen. It was thought electromagnetic attacks were old-school technology, causing researchers to fail to pursue the field. A couple of technical papers Jemma read in the guild's library caused her to review physics. The magnetic bomb launchers would create a powerful electromagnetic pulse that would create backflow loops in any unprotected circuitry, causing burnout and electrical failures. Turk's mystery inventor also delivered a means to harden all the station's electronics against these weapons. The miners, frustrated by Allison's unprovoked attack, had mobilized and turned out in droves to help. The final defensive masterpiece was the massive protective shield for the station. The shield required diamonds, which drained the diamond crystal clusters the guild had slowly amassed over the years. The station could now handle a battleship Ion cannon attack in a defensive mode that few colonies could defend against.

Jemma was actively involved in outfitting Mining Sled Two for the upcoming summer mining expedition designed for the youth. She ingeniously integrated lasers at the sled's forefront, ensuring they blended seamlessly with the design for discretion. Additionally, she mounted two external tubes along the shuttle's sides, enhancing its operational capabilities.

Together with the youth, Jemma oversaw the installation of four new engine pods, significantly boosting the reactor's power output. The team then equipped the sled with magnetic rail guns, positioned for both forward and reverse firing, and attached them to the tubes for strategic flexibility.

A more formidable rail gun was mounted atop the sled, an offensive powerhouse requiring large ammunition slugs, limiting the ship to a total of seven due to their size. On the sled's top leeward side, they installed a concealed plasma cannon capable of launching powerful plasma balls similar to those used by the station's defenses. This cannon was cleverly designed to rise through a port for engagement.

Drawing inspiration from a battlecruiser design obtained during a scan of Twitch, Jemma also incorporated sand canister defenses into the sled. However, she innovated by using ground metallic asteroid fragments mixed with bright pink fluorescent powder, enhancing both the defensive capabilities and the visual impact of the sled's countermeasures.

The group of young adults was actively engaged in transferring their supplies and personal belongings into the shuttle, where they had ingeniously reconfigured the space to include four private cubicles. This modification allowed them moments of solitude whenever they desired. The day prior, they had already stocked the sled with food supplies, not forgetting to include a selection of their most loved candies. Jemma contributed to their preparations by enhancing their rock cutters and tweaking them to operate with greater heat and efficiency.

Energized by a brainstorming session, the youth revisited some of Adrian's utility drones they had previously set aside, such as the chemical cutter. They planned to upgrade these drones with advanced components and program them to perform with a new level of agility and innovation.

Furthermore, they constructed a new trailer for the sled, specifically designed to carry extra provisions, notably oxygen and water. Recognizing the unmatched value of a hot shower after a long shift, they aimed to address the traditional sled's limitations on water supplies, which typically necessitated recycling. This storage trailer, intended as a temporary supply holder, would be filled with mined ore once its initial contents were depleted. Unlike the conventional tow line method, this trailer was ingeniously crafted to attach directly beneath the sled during transit, streamlining their operations.

They also assembled more database information on the asteroid field. The new information included additional age dating, including several rhenium-osmium data points on the older units encountered in the belt. The Nickel-Iron core material was logged based on specific gravity and magnetic polarity. Sue's incredible inspirations include logging hydrocarbon compositions, age, and geochemical fingerprinting. These would link similar occurrences in asteroids containing old petroleum systems. They had also typed structural elements of the near-surface sedimentary basins, knowing thrust belts and magmatic arches were special in identifying subduction and potential for continental keel formation. They had amassed a significant database, and Jemma had gone deep into correlating components of the database, looking for high-probability matches of asteroids. Her mapping was already beginning to reveal patterns not predicted in any of the hundreds of academic papers related to mining deposits of the asteroid belt.

Finally, the family gathered at an old-style wooden picnic table. They reviewed plans, goals, check-in frequency, emergency procedures, and flight paths. The mining supper, as it was called, was a family tradition before the summer mining season.

John was sitting beside Tony, chatting with John's dad. "Dad, we also intend to investigate leads from our asteroid reconstruction project. Of course, before we embark on any investigation, we will provide detailed primary and secondary flight plans." John thought his Dad would react to them moving away from the primary claim area.

"First, you will complete your claim assessment and begin prospecting – right?"

Tony jumped in. "Correct. We assessed the survey you completed when preparing to cut Uncle Adrian's sled out. That survey has pointed out some interesting anomalies within thirty meters of the surface. We intend to survey the rest of the asteroid with one team while the other drills exploration holes for the anomalies. We will use conventional drilling to evaluate the cuttings for evidence of Kimberlite magmas. If we encounter diamond or indicator minerals or experience a bit-drop, we will send a probe to evaluate the potential."

As if she had rehearsed her conversation with her father, Tanya said, "We intend to have the initial survey completed within three weeks, and leads and prospects prioritized within a week after we complete the survey. We will use a similar grid used by Uncle Adrian except expanded and include an ultra-low and higher upper frequency. We will also spread the source to signal pairs and consider a random source grid. We believe this will solve the problem of poor-quality imaging from the original survey. Jemma will process the data on the fly. She is, after all, a significant improvement of processing power for a mining sled."

Jemma twitched at the sound of her name. "We will have a great picture of the Asteroid to nearly its core." She smiled as only a daughter can to her father.

All of the youth stood to help move the leftovers to the Kitchen. Tony's mom, Chivaree, stood and said, "Now the business meeting seems to have covered the basis. It is time for our lunch. I need help to carry the leftovers from the table."

The parents had planned to relocate to Asteroid Sixteen two weeks following the youth's landing on the asteroid. Adrian's offer to escort the young explorers to their asteroid claim was met with a level of enthusiasm akin to that of high school students being chaperoned to school by their parents—polite but unmistakably lukewarm. Following a lavish lunch that marked the final meal in their current homes, the youth retreated to the sled for their customary post-meal nap while the parents engaged in their traditional game of hearts. This grand feast and the following rituals were steeped in a long-standing tradition cherished by the families, serving as a fitting farewell to their familiar lives.

Inside Sled Two, John was lying in his bunk in his cubical and hit the com. He was a loving brother to a sister in need. "Sue, I never thought I would say this, but thank you for being a great sister to us. I would like to have another sister and am so happy you have chosen to come with us. Without you, we would have had a big hole."

"Thanks, John. I am happy to be part of this family."

John could hear the slight sniffles that were occurring less frequently. Sue missed her parents.

Chapter 19
Cirrus Minor - Exploration - Station-Next Morning

Mining an asteroid requires exact mapping and position as it travels in its established orbit. Mining teams consistently outperformed individuals for several reasons. The first was companionship. Companionship requires communication. The second was planning. A prescribed plan allows the work to be approached from a logical, well-thought-out timetable. Discussion of the program always included safety discussions. Miners would stop before a task and review safety. They worked as a team and were naturally focused on the same goal. While individuals were safe, there was nothing better than a constant reminder. The expression there is safety in numbers is accurate.

Sled two was perched at the air barrier with the sled in takeoff position. Tanya was piloting with Sue in the co-pilot's seat. Tony was at the communications terminal, and John was in the sled's main living area at the sensors. Jemma had lain down in a mining reclining chair, appearing to sleep in a dead dog position, feet up and on her back. She was projecting a mini hologram of the sensors that John was watching.

"Control, Sled 8082 requesting departure, over."

"Sled 8082, an affirmative. The lane is yours; good luck and happy hunting! Please double-check outbound flight plans. We verify your special mine leg secured flight plans are secured."

Miners filed outbound flight plans so the sled control could control the spacing of the sleds outbound if needed. The miners had inbound and outbound lanes. Beyond these, miners would venture to their claim in a circuitous route, a practice from old Earth mining. Miners were challenging targets, but they could be followed. New mining laws protected active claims, but other areas were open. A hot claim with a significant discovery could become a Wild West gold camp of the 1880s on Earth in North America – crowded.

The problem was not other miners. Earth produced glory seekers who would come from Earth seeking fortune and fame. They would eventually fail. Those who failed their initial mining attempts frequently shifted to piracy, lacking any other viable skills. Without funds or backers to return to Earth, they made a temporary impact before fading into obscurity. Fortunately, in contrast to the goldfields of California, the presence of these glory seekers and their corrupt ways didn't enhance the prosperity of the prosperous Cirrus Minor Asteroid Field. It was the miners, adhering to a shared code of morals and ethics, who ultimately dominated this space.

"Sled 8082 to Control, affirmative over."

"Control to sled 8082 good to go." The controller said.

As a skilled sled driver, Tanya pulled the sled out through the air space interface and into the lane. Sue went through another checklist and pulled sled 2's utility trailer into position under the sled.

John and Jemma initiated their tests with the passive sensors and then proceeded to evaluate the active sensors. The standard configuration for most sleds included a radar and an automatic ship identification system. However, Sled 8082 was exceptionally outfitted, its sensor capabilities far surpassing the norm. Its active sensors covered the full spectrum of light energy, equipping it with a sensor range comparable to that of a battleship. Jemma had enhanced the radar's reach, doubling its range, and added a Doppler radar for precise close-range tracking. The sled's compact lasers were integrated with the Doppler system for heightened accuracy. Furthermore, Jemma incorporated a magnetic-telluride sensor, adept at detecting low-level electrical signals from ships concealing themselves within asteroid clusters or stealthily following the sled.

Jemma's previous experience as an AI fighter included secretly hacking into Allison's Mantle Works computers due to her curiosity. She had observed a great deal of technology. As a result, the sled was packed with high-end equipment printed and assembled in the maintenance hangar. Access to the mining guild printers also helped her get a hold of some of the more rigid metals required for some of the unique component requirements. Jemma's six-digit paw conversion was helpful and significantly sped the process up.

Jemma woke up and approached John. "Can we make a slight detour to conduct a small weapons test? As you know, it is my nature to have offensive and defensive capabilities. This ship now has both and, in a pinch, should be able to tackle a destroyer or more. After Allison Jiggs's recent incursion into the station, I learned that this sled needs further additions that could be either offensive or defensive. I also borrowed some ideas from a small conversation with Twitch Isaac Bauer's robotic pet."

"I will ask the group. Will it be like a fireworks show?"

"That's why I desire to take a small detour to get out of the common travel route and into some small dense asteroids. Light show, yes. Fireworks not in the sense of a celebration."

John opened up the communicator. "Hey, group, Jemma wants to show us some of this ship's capabilities and treat us to a light show. It will require making a small detour into a group of small asteroids about 10000 klicks off our current course. I am guessing a delay of three hours from our current schedule, which we are ahead of. I need everyone's concurrence."

"I'm in." Sue quickly replied.

Tanya was second, "Me too - a light show - this will be interesting."

"I was just getting bored driving this ole sled," Tony replied. Tony was in the pilot seat after Tanya rotated out. The ship was on autopilot, so there was little to do.

"I am sending a new flight path variation," Jemma announced.

"I will send a message to Cirrus Minor Base to let them know we will be dark for about three hours," Sue said from the communications council.

"I have the path plotted. Holy Smokes! Are you sure there are a lot of small asteroids along the flight path!"

"We will be clearing a path as our test of defensive and offensive capabilities," Jemma replied.

"I am so excited. This is what it is about! A cowboy ride through the trees. Ye-ha!" Tony stated

"Settle down, pilot!" Tanya spoke up from the co-pilot's seat. "Are we activating the Plasma Cannon?"

"Later in the test, I will run the defensive weapons."

"OK," in unison.

"John, go dark, deploying the ship for passive sensing." Jemma requested.

"Tony, here we have about 15 minutes before encountering the first asteroid debris."

"10-4, 14.25 minutes," Tony had become somber and professional. "I am ready to fly. Copilot, let us check that list again."

"Passive is picking up a large signature inbound," John said, slightly surprised.

"Correct. That is a large resupply freighter. It is an old model. It leaks electromagnetic like a sieve. It is fifty thousand clicks." Jemma stood up and stretched, acting like she was sniffing for something in the air.

"Fifty thousand clicks?" Tanya asked

"That is my best estimate. We will know exactly when we go to active sensors." Jemma sleepily replied.

A few minutes later, Jemma announced active sensors. Immediately, Sue informed the group, "Debris in four minutes."

John activated the debris shield at the two-minute mark. Jemma then powered up the defensive lasers and the Alexandrite Lasers mounted at the ship's forefront. "Our initial trial involves piercing a path through a debris field, specifically an S-shaped route. We'll begin with the front Alexandrite Lasers, known for their superior power. This will allow us to refine the technique for focusing the lasers individually or in groups. The programmed cutting area is set to be roughly one and a half times larger than the ship in both width and height. Our opening should form about 5000 kilometers ahead of us at a speed measured in millisecond kilometers. Should the window not form as anticipated, we'll incorporate auxiliary lasers and the plasma cannon. Please remember the last resort is the ship's exterior shield."

"Good, there is the plan," Tony conned in. "It should be fun to fly. Tanya has projected a flight course line that Sue and I can trace in a ship with unproven weapons and shields! A good test. In twenty, nineteen,...."

"I recommend recording the composition of the material we cut through. We currently have no information related to this debris field in our database." Sue requested.

Tanya immediately responded from the sensor panel. I have already programmed the sensors to collect this information. I will execute the sweep now".

Twin purple beams extended from the ship, slicing through the vast debris field. Thanks to Jemma's precise aperture settings, the debris directly in front of the vessel was instantly vaporized, carving a path through the dense asteroid cluster. As the ship navigated through, the purple beams adjusted to create a curved trajectory through the debris. Tony, gripping the control stick with tense determination, maintained the set course. Sue, possessing superior piloting skills, was ready in the co-pilot's seat and prepared to step in if needed. The crew was learning to rely on each other's abilities. Being a passenger at a velocity of 1500 kilometers per second required a deep trust in the pilot's capabilities, given the slim margin for error. Little did they know Jemma was ready to take over instantly. There was no need, however, as Tony was glued to the course. No other weapons were needed as the Alexandrite Lasers were potent and cut a path through the dense field without issues. Even the highest density portion of the field immediately vaporized under the lasers. The external debris force field was activated with only minor asteroid debris fragments. The S-shaped cut went very well. The sled exited the debris fields, dragging a fine mist of debris.

"Okay, Tony, drop the speed forty-five percent," Jemma asked.

"Forty-five, Yai," Tony repeated.

"Tanya, I have spotted two large rock masses bearing 270 horizontal and 14 degrees. Please confirm."

"Confirmed Nickel iron composition density 3.5 grams per cubic centimeter."

"Firing main plasma cannon at rock labeled one," Jemma responded after a moment of silence.

A green beam emerged from the ship and vaporized the asteroid. A whirring was heard in the boat. Jemma had activated her second plasma cannon, which fit into the leeward side of the upper haul. The artillery and its mount were in a particular receptacle on the upper section of the sled. This cannon looked like a five-inch turret on an old earth destroyer used on the sea. This was a new design. It fired a series of green plasma balls at the second rock. Which abruptly disappeared in a massive bright explosion.

A few seconds after darkness returned, Jemma quietly said, "Oops!"

John jumped up and went to the screen. "Holy Fireworks, Batman, what was that, Jemma? That cut rock number two like butter and then continued."

"Yes, it appears only one ball was required. A miscalculation on my end." John was surprised by this answer. "Jemma, tell me about the miscalculation?"

Jemma turned, looked at John, and sat down. "Not a miscalculation, but the test was in the thirty-fifth percentile on a probability curve. Due to some of the component uncertainties, we could only provide a probabilistic estimate of the power. In this case, nearly everything went in our favor." The dog provided a toothy smile of apology.

Sue was quick as lightning, "Jemma, can you tell me what a P1 outcome would look like?"

She turned her head, tail wagging. "That is the right question! I will set up a physics class immersion session and run through the physics in a few days. That topic flips this dog's switch." She walked to Sue and sat down. "Now for some more good stuff."

She nodded her head toward Tanya. She raised her hand to the pink color button with a can etched into the pink metal covering.

The dog continued, "Now, to test the anti-missile system we borrowed from Isaac. Two bumps were felt in the ship as two small canisters were ejected in front of the boat. They moved to about 15 kilometers and burst with a spectacular metallic pink cloud forming.

"That, my friends, is an anti-missile screen and a deterrent against fighters. The cloud is comprised of recycled metal shavings from 3D printing operations, making it highly abrasive. For visibility in this test, it's been tinted with pink powder. The rest of the canisters contain lithium, detectable by our mass spectrometer. Although missile attacks on a mining sled are unlikely, it's wise to stay prepared. With that addressed, we can proceed with our original navigation plan. Remember, your father instructed me to ensure your safety while allowing you the freedom to explore. Pilot, I've forwarded a suggested route for your approval. Next, we'll need to carve another corridor through the debris field." The captain was always the pilot in the miner's code.

"Sue will be driving this time," Tony replied, still wrenching his hands.

"I will practice with this new sensor array," Tanya added.

John rose and started to walk to the communication council. "I will prepare a note to home base that we are back on our flight path."

Five hours later, the youth reached the claim. They stopped and pulled a crazy Ivan loop to ensure they were not followed. Tony renamed the old Russian submariner's trick to Crazy Adrian. The maneuver was designed to shake twenty-first-century enemy submarines off a submarine's stern. Turk insisted they pull out all the stops to ensure they were not followed. They had not been. Jemma's enhanced passive sensors were enough to tell them that. It was time to land on the claim that had already changed their lives.

They entered the claim code to validate their right to work the claim. The mining claim buoy signaled to the Cirrus Minor Station Mining Guild station that a valid access code had been entered and by what sled. The mining claim was still secret, but the claim designation was switched from newly claimed to active. Tanya was piloting with Sue in the copilot seat as they dropped the auxiliary sled down first, then rotated the mining sled and landed on the asteroid. Previous work conducted by their fathers was evident once they had landed but hidden from the view on an overhead pass. They changed to their mining suits, including their hardcover shells, and explored the cuts where the vug had been removed. Upon return, John and Sue pulled out the survey bots and a quad-track mining tractor on a prepared sled. They would use this to expand Turk and Adrian's survey to identify other diamond vaults.

Meanwhile, Tony and Tanya rolled out another tractor, this time attaching a drill to probe the previously pinpointed anomalies. They also fitted it with a dozer blade to clear away the debris left behind by their fathers. Once both teams had finished prepping the equipment and performed their final checks, they made their way back to the sled for the customary meal of good fortune, a cherished miners' tradition. Upon entering the sled, they were greeted by a dog, now sporting a newly tailored, hard-shelled mining suit designed specifically for canines. Jemma was in the process of putting together the helmet she had crafted on the 3-D printer.

John, surprised, looked at Jemma and smiled, "I have seen it all: a space dog in a dog-mining hard shell. Do you mime too?"

Tony broke out laughing, seeing the dog. "Holy Cat Chaser, this is great. I am giving up my day job and hiring a herd of you to mine for me."

Sue scratched Jemma's exposed head and said, "Smart dog," and Tanya followed suit.

"I am impressed you seem to be developing a creative side. I never dreamed of seeing a dog or robot in a spacesuit!"

Tony said, "Jemma, we could start a new line of dog space suits. We will need a tether, however. I can see throwing a tennis ball in low gravity and watching your dog go after the ball jumping into deep space because it does not understand the effects of low gravity ."

"Funny, Tony," Sue quipped.

"Tanya wants help with this processor?" Sue moved over to help Tanya. John set the table. As Tony continued to inspect Jemma's new suit."

Jemma looked at John. "I can only do so much with remote sensing. I can do so much more with this. Technically, I don't need a spacesuit, but I have some great features that will be useful.."

Tony rolled his eyes. "A cowboy dog with a spacesuit will travel." He got up to help the other three with dinner.

After dinner, Tony was sitting quietly with a puzzled expression on his face.

Sue looked at him and asked, "Tony, what's up?"

Tony turned to Sue. "Why did my Dad drive his sled into this fissure?"

"Crazy Adrian said he had a map, "Tanya returned. All the kids were used to calling Adrian crazy Adrian for a lot of unexplainable behavior he was known for.

"He gave me a copy of his map." He flipped it to the hologram from his wrist computer.

The map showed a set of asteroids with a chaotic path and notes added to the side. It was a flat map, and each Asteroid had a series of notes related to size and sometimes composition. Occasionally, an attractive shape was designated with a drawing, such as an Asteroid they had landed on. It had a keyhole-looking feature with the flight path going through the keyhole, lining up for the next turn, which would go to the x marked on the map. Tony laid out the three-dimensional map of the local asteroids and tagged those with unique compositions from the database they had compiled. He could not correlate enough points to rectify it with the 3D map. Sue noticed three old-style sextant measurements on stars that she could interpret. These changed the map's orientation, which helped focus five of the asteroids, but Adrian had already flown those according to notes he had recorded in the flight navigation data for his last flight. It demonstrated some validation of the maps' authenticity but did not validate the end purpose of the treasure hunter's map.

"Might give us another piece of information and help us prioritize asteroids we kick out as potentially containing mantle keel material," John added.

Tony, the skeptic, then added, "Never been one for Dad's treasure maps. This one, however, did yield mantle keel information and Draconia. He highlighted the peridotite and eclogite asteroids on the map from posts or the physical database. More than 50% of the mapped asteroids had matches. That was beyond coincidence.

"Did Uncle Adrian make a wrong turn, and the keyhole is nearby?" Sue asked.

"Possibly, but he is an excellent navigator as long as he is not distracted by a treasure map. Then all bets are off!" Tony asserted.

"We have time for a brainstorm if you are agreeable," Sue piped in. "I would like to look at structural relationships in basin-type asteroids linking sediment age and timing of compression in the thrust fold belts. Maybe we can separate accretionary prisms from thin skin fold belts. I think it would tighten our classification scheme."

"Yes, and typing petroleum system information might help determine which asteroids of these thrust belts are tied together," Tanya said, moving to the immersion headgear storage.

"I'm in." John followed Tanya to the storage cabinet. Soon, the four were immersed, and the 3D halo map rapidly changed to reflect common asteroids. They had come a long way, which was reflected on the map. They had not attempted a reconstruction yet, but that was coming shortly. Clusters of similar blocks were evident, which would cluster tightly in the animated rebuild of the lost planet Cirrus Minor. The most important were the continental shield areas and subducted blocks to help focus them on the mantle keels. Jemma multitasked in the brainstorm and donned her helmet to test her new hardshell mining dog suit locally. She clipped the final latch and reviewed her checklist, including a pressure test. At the same time, she did not need a spacesuit but had significant sensors built into the hard shell. This included a small plasma torch and a couple of lasers. She quietly exited the airlock, testing her shuffle step to keep herself from self-ejecting from the asteroid by a false move. She found she could walk fairly normally but had to be careful. She had a small rocket pack built into the suit as a precaution. Jemma had split her conscience to manage the immersion session while she went to test her hard suit.

The artificial intelligence was curious about the gap Adrian had flown into, too. It was a pitch-black hole in the rock without evidence of an exit. She flipped her suit's lights on, which were absorbed by the dark opening. The dark eclogites did not help. She continued into the black cavern but did not see anything of interest. Jemma noted that neither of the youth's fathers had entered the gap. Jemma was echo-sounding the cavern by leaving a ground thumper near the opening of the grotto. As she proceeded forward, the little thumper kept pace. The space-suited dog launched three drones to capture the timed arrival signals through the rocks on the sides and ceilings. She walked in a box-like pattern while recording the thumper arrivals. This process could also identify any additional near-surface vuggy caverns in the eclogites. She processed the high-resolution data and sent it to the halo projector before her crewmates.

"Jemma, Where are you?" John asked as he put on his hard suit bottoms.

"I am testing my suit and its capabilities. Ships sensors could not penetrate the eclogite when I scanned the mapped Keyhole feature. Since I needed to test equipment, this seemed logical. The halo is projecting the edge of the cavern. I am also projecting my suit's external cameras. Additional vug-like anomalies like your father's recovered will show on further processing. Since this area will likely find another large diamond vug, we must survey this as part of our mission objectives."

"This area is all in our highest priority search area," John smiled. His fingers played across his wrist computer. There!" A window containing a terrain elevation model depicting a 3-D surface map with a colored grid across the top, "The pattern is hot color a priority to lower priority winter colors. Jemma's position is the moving golden triangle.

Tony lit up, still in his immersion gear. "Hey, I can see all around you by turning my head. This is cool, but it looks pretty dark in there. Thought about a light drone?"

"Yes, but any vibration, even the wisk of a jet, could impact the cavern survey. It is a bit old school." She concluded.

The miners nodded their agreement because they knew they would soon do the same in their survey operations. John marveled that the dog had been learning the modern mining techniques, adding a few new twists. He was still wary about the significant offensive upgrades Jemma had made. She was, after all, unconstrained by robotic laws, was born into a space fighter, and was growing in potential.

Then he asked the all-important question? *What constrained her?* She would not physically implode if she violated a coveted law. So why tempt faith? That caused him to review what little he knew about Jemma. She tried to commit suicide when asked to kill. She helped whom she could but hid from Allison Jigs and the goons. She had restrained herself or played an authentic game of not tearing a guard's throat out. She was a teacher and a student and possibly a conscience. Was she living? That would have to wait. He wanted to see the rest of the cavern, so he unsuited and joined the rest. He knew Jemma was growing fast but had a heart of gold. Funny, he thought of that story set on Earth about a girl in Kansas caught in a Land of Oz with a brave Tin Man with a heart as big as gold. Was he the strawman or the Lion? John focused as he put his headset on.

Jemma was continuing her box pattern. She would do that for another four stations. They could see the cavern closing but were getting mixed signals at the tip of it. Suddenly, a bright lightning bolt flashed from the floor to the ceiling. Jemma's center-upper drone and one drone by the side of the cavern ceased functioning and dropped, making no sound in the vacuum as they tumbled slowly down. The team also noted the drone's failure immediately after the flash.

"Jemma, put that area off-limits, please. What is interfering with those electronics? You could be compromised too if you are not careful." Sue sounded.

"That's true, but they are not protected against electromagnetic radiation. I am, but I will go slow."

John was already up again, putting on his hard suit. He led the entire team, who were now suiting up to help their teammate if needed.

Jemma was moving slowly and monitoring the electromagnetic wavelengths carefully. She could see the occasional E.M. bump, but so far, nothing dangerous, and her downed drones were visible with her bright suit light. The third drone, she recalled, plotted a careful course back to the cave opening.

John and Tony had mounted tractors. Tanya and Sue were on their tails into the second tractor. They were ten meters from the ship and moving to the cavern.

Jemma shut the survey down with one last vibration sequence she recorded at her post. Then, she completed a 360° scan of her surroundings. Sue was monitoring Jemma's video feed. She abruptly alerted Jemma of an anomaly. "Jemma, STOP, turnback 15°, please."

Jemma complied and stopped, turning 15 degrees to her right. Nothing looked abnormal in the darkness of the rough cavern floor at first glance.

Then Tanya excitedly said, "Oh, that was a sharp observation, Mags. Zoom in and down two meters from the center and three meters left again from the center." The view zoomed as directed to image a slight dark shadow with a rounded appearance. You did say Uncle Adrian or my dad did not come in here?"

Jemma responded, "There are no tracks or evidence of disturbance in this cavern for at least decades."

"Why is there something round in the shadows?" Tanya had stopped behind Jemma's position, with Tony parking beside their buggy.

John rotated the tractor floodlights toward the end of the cavern. They could see several rectangular and rod-shaped objects with long shadows being cast by them across the end of the cavern.

"I did not see a record of anyone mining this asteroid in Dad's claim report. Do those look natural?" Tanya questioned.

Jemma moved forward, scanning with every bandwidth she had while further examining the rounded object. The object Tanya found was registered as having a metallic content of aluminum and copper. She slowed, alerted that composition was a refined ore with limited potential known sources - man. She had also conducted a search related to the claim. She validated the Father's claim that no available guild member had ever claimed this asteroid or any near it. The lightning bolt and the knowledge that no known group had ever been on this meteor led to two possibilities. 1) pirates, 2) since no living carbon-based forms were present, booby traps could be present. Jemma also noted an elevated carbon source in the direction.

The space-suited quadruped then stopped at the rounded object identified by the young ladies. It was an old can that was empty with no detectable residue. She stopped and cautiously scanned the area, then the can. It looked new, except the labels were not a language she had learned. There was a broken bottle under the can that she also scanned. Again, the text and picture on the bottle were unblemished, but the contents were long gone. Next, she noticed a large piece of rigid environmental plastic in a low spot, but the composition was incorrect. It was made of a silicon substance. It looked like it could have been a barrier to an atmosphere.

She continued a few feet further when she found an old grey waxy mummified skeleton holding something refracting bits of light directly from its suit. She cut her video feed, puzzled by that reaction. Was she hiding something from John? Yes temporarily. She needed to process this.

She looked down at the old preserved being. The biped skeleton did not have features entirely consistent with those of a human. She looked at the old mummified being and realized it clutched a Draconia crystal peacefully in its hands. Jemma now had an abstract moral question to resolve. If she removed the Draconia, she might be desecrating a corpse. Perhaps this was a burial ground or some horrible way to strand an enemy. Jemma needed specialized, trained abstract thinkers, one or all of the humans. She had not cleared the area as safe to work, either. She had an idea about the lightning strike, the wobbling electromagnetic, and the Draconia in the individual's hand. They could test her hypothesis with the Draconia in the lab. She pulled a special container out and gently set it beside the extinct body. With her paws converted to her six-digit hands, the robot AI reached over the body and gently lifted the crystal from the mummy's hand. The robot had previously scanned the body for a dead man switch, which was clean. The Draconia fit nicely into the container, which was replaced in its receptacle. The dog turned around to face the opening of the cavern. The lights from the tractors were bright. She switched her full suit-feed back on immediately receiving a call from John.

"Jemma, good you moved. Are you ok?" John rambled. "We are beginning to construct a sling cradle to come to get you. Wait there. OK," a millisecond pause. "OK?"

"I have uncovered an anomaly that will change the group's plans. We need to return to the sled and conduct a simple analysis. Everyone's safety is a concern." She carefully started to retrace her steps.

John grasped the situation instantly. "Tony, Tanya, and Sue, we must pause our current tasks to fully grasp this anomaly! Let's move."

No one contested. After all, they were miners adhering to an unwavering code where safety took precedence, and the mining sled was their haven. Without delay, the team boarded the sleds, with Sue and Tanya setting off first, followed by John and Tony, who lingered for Jemma.

Once aboard the vessel, the quartet shed their upper protective gear, eagerly anticipating Jemma's explanation for her concern over their imminent danger. Jemma entered, her helmet now securely attached to her suit.

"Well, Tony asked. "What about that "plasma bolt"? It looked like a simple lightning bolt to me!"

Jemma reached into her hard shells storage bin and pulled the Draconia crystal out of the sealed plastic container. An audible gasp filled the room.

"Did you find Draconia? Tony blurted out. "You go out for a little stroll and return with a small fortune?"

John asked, "Jemma, did you pick this up in the cave?

"Yes, I did."

Tanya: "How was it embedded in the wall?

It was not in the wall, Tanya. It was close to the ground. This should be considered a biohazard until cleared!"

"What do you mean close to the ground? Was it magically floating?" Tony jumped in.

"No, it was in a corpse's hand, hence the quarantine protection. The cave's back portion was sealed at some point and contained an atmosphere. The corpse is a victim of atmospheric starvation followed by long-term exposure to space. The enclosure is long gone. I have a hologram of what I recorded."

"Holy ghosts, Jemma, maybe this place is haunted," Tony said, thinking hard about the discovery.

Sue stepped in. "Yes, I would like to see?"

The ghoulish image of the dead hominid scene drew gasps. The mummified corpse was lying on its side. It was a biped dressed in some suit. Its helmet was lying neatly beside it. It was a design the youth had never seen. Its arms were folded across its chest. The video showed the glimmer of Draconia it held in its hands. The facial features were just a little off. The creature had a massive eyebrow but no hair on its face. It had a peaceful serenity about it. The animal had died at rest long before the curtain deteriorated enough to release the atmosphere.

That was clear to the group looking at the scene. The corpse did not indicate sudden compression due to loss of atmosphere. Everything was in an environmental shell except the hominid's head. Jemma finally showed a picture from the ceiling drone. The photo showed a clear pattern of tiny twinkling lights arrayed in a clover pattern. "This is the floor of the cavern before the energy pulse." The following photo was the same, except the sparkling spheres were now steady and bright.

"These are growing in intensity, and when it released the cavern, they released their energy. After the explosion of plasma, the cavern returned to normal. I believe we will find a Draconia crystal at every one of those point sources. I will test this crystal for a latent charge, which will help with an interesting hypothesis." She moved the box containing the crystal under the environmental protective lab hood. She produced a simple testing kit comprising nothing but a light bulb on a stick. She touched the metal end of the crystal, and the light bulb began to glow. The light grew in brightness until it popped, spraying a shower of thin glass against the thick transparent wall.

Tony understood the meaning of the test. "Jemma, that crystal is carrying a hefty charge."

Jemma looked up at Tony and smiled her doggy grin, "Correct, but it is gathering the charge. This pattern seems designed like a dream weaver, except it collects an energy charge."

Tanya weighed in. "Right, it is a battery, and this person found out how to charge it?"

"I believe that is correct. This appears to be like a battery. It ran equipment for this individual user. The battery has been left in the charge position."

John joined. "Looks like it is in a charge-and-spill mood. We happened to visit the site, but it was nearly full. We witness peak capacity."

Jemma moved ultra-slowly on the cavern floor before the plasma was released. As the plasma was building, the image displayed several small pinpoint light sources scattered across the back of the cave. They were formed in a specific pattern, like small groups of six-pronged stars.

Sue then spoke. "We have a genuine mystery to solve. We should investigate this if it is safe before we go mining."

John spoke," I agree this should be a priority for us to investigate in the morning. I noticed the cans and agree this must have been a sealed closure. This individual is not wearing a hard-shell environmental suit. It needs investigation."

"At a minimum, we need to have a burial for this poor soul." Then Tanya added, "This must have happened a long time ago. His clothes are a style I have never seen." Tanya added.

Tony, whose imagination had gone wild, "I bet this is an old pirate hang out! There is a chest of gold and diamonds in the back. We know there is Draconia. What are we waiting for?"

Sue rolled her eyes. "The Draconia is interesting, but I doubt this was a pirate. I vote with John. The first thing we should solve is this mystery, but not disturb the site too much."

John stood up and yawned. It was getting late. "I have enough to sleep on. I think we have some work to do before we go out. What are the rules about finding a corpse on your mining claim?"

Tony stood up and said, "What about the prize money for finding the pirate base?"

Tanya put the dampers on the conversation. "What about the implications of this being an alien's mining site, or is this a sacred and protected grave?"

They all agreed and went to Bed with Tony, making guttural ghost-moaning sounds.

Chapter 20
Cirrus Minor - Ray Gunn - Asteroid 16-Isaacs Apartment

The pleasures of life are sometimes that little written word on a coffee cup like "Smile." Eloquent and straightforward, but enough to change a life. Ray Gunn sold pottery manufactured on Earth. It was in high demand and fetched high prices in the Cirrus system. Some thought it was just a silly gimmick. It was so popular that a small group of shoppers would arrive at a business early to see what the Ray Gunn people had created. Some would develop online companies to buy reasonably priced pottery and sell it for an arm and a leg. It was big business. However, having a mug or a plate talking to you can sometimes perplex you. Thanks, Ray Gunn. You saved someone's life today.
*–HAPPY_*

Isaac was sitting munching on his favorite brand of dill pickle. It was a giant pickle of which he had consumed about half. "When do you think Allison will try to assert herself here?" He asked Mackena on the wrist communicator.

"I'll be walking in the door in two seconds. Then we have the day to chat".

She had just transitioned the apartment's decor to a summer motif. The apartment now gleamed with bright summer colors and was dominated by a large halo picture of a hillside blanketed in wildflowers, which changed as the seasons shifted. Currently, after a recent shower, the apartment was filled with the scent of damp earth, reminiscent of the aftermath of a storm.

Mackena walked in with Twitch, towing a small gravity sled filled with boxes labeled JJSpace. Isaac knew Mackena's great weakness was pottery from this store. The shop owners knew this and carefully tracked her material purchases made by an artist named Ray Gunn. She could not buy these on Cirrus Major due to the required security around the Trex's construction. She was now making up for being deprived of this opportunity for so long. The shop owner was thrilled when she showed up. No one else in the commercial sector of the Asteroid 16 colony had Ray Gunn's product. Isaac chuckled as he looked at his white porcelain mug with a word etched in black- happy. He smiled, thinking, Damn, the cup is talking to me.

Mackena excitedly started removing new pieces from their boxes and replacing several old mugs with new ones. Smile, look, please, thanks, fill me, mum, jump, and run fast were a few labels on the mugs she replaced. Then she put down a new white purse with the word 'stuff' written in black.

Isaac stood and stated, "Are you kidding me? Where are the rest of the sentences? Verbs and more nouns would allow us to give up most forms of communication and use the mugs."

Mackena kissed him on the cheek, "Back to pickles, I see. I did not know if you would find them."

"Get me a mug that says pickle," Isaac replied, thinking he was coy.

Mackena moved a few boxes, pulled out an enormous container, and pulled out a vase with a sealable lid that said 'Pickle' emblazoned in bold black letters. "Thought you'd never ask."

"Are you for real?"

"You think you can out-think me? Wait until you see these birdhouses. "

"Gasp, and you are smarter than I am, so no contest," Isaac said dejectedly. "The answer is simple, Mackena. What birdbrain would build a birdhouse for a bird in an airless environment? I am in trouble if I hear a bird singing in space." Isaac smiled. He switched subjects rapidly. "What do you think Allison's about to do now that she has figured out I have departed the company?"

"How do you know that?"

"Well, she visited Cirrus Minor Station. She brought two destroyers. When she met the Guild Master, she burned the station's defensive weaponry. Then she visited our new friend Turk, who went toe to toe with her. Nearly scared the pee out of the guild master. Her goon Chadoom was next door holding Adrian in the air as his men tore his maintenance bay apart, looking for our little prize, which is on the last leg to Earth."

"What about that nice dog, um, Jemma," Mackena asked.

"Like Twitch, Jemma hated him, but unlike Twitch, that dog obeyed Turk's command to place. The guild master was quite talkative."

Intrigued, Mackena asked, "Do we need to help the guild shore up their defenses?"

"That is where it gets weird." Isaac looked puzzled while he momentarily thought. "Allison pissed off the miners. Turk told the guild master that a friend had already designed new defenses for the station but needed a workforce to install new 3-D-printed weapon parts. They now have more capability than we do. And they have a canister defense. Where did that come from?"

"We detected no scans of our ship while it was in orbit. Do we have a mole?" Mackena asked concerned.

"I don't think so. These are vastly improved. The guild master thought we did it. I told him no, but if he had the blueprints, we could market them. He agreed and sent them. Since they do not have a copyright, I entered into a contract with the guild to copyright and patent them, and we market and sell them, retaining forty percent. He agreed. Since I can't make hide nor hair of these, I thought I would leave them in your hands. Look at the alexandrite laser. Impressive."

"Sounds interesting. I will get the engineers on it. Did you see these sand canisters have pink powder or lithium as a marker? Who would do that besides a lady? A genius lady. What degrees do Turk and Adrian's wives have?" Mackena took some time to look at the minor details of the plan.

Her expression shifted dramatically as she went on, "These are astonishingly well-documented. The 3-D printer settings are integrated right into the master code. Who would design a weapon with an option for a pink powder cloud? These are exceptional – I doubt there's anyone nearby capable of such craftsmanship. You'll need to search for a genius with an exceptionally high IQ who has a fondness for pink and is likely female, given our superiority within the species. I'd recommend combing through the station's records for someone with the skills of an advanced master builder or someone of equivalent talent, likely operating incognito."

"Interesting you suggested this path, Mackena," Isaac said quizzically. He continued, "I asked the Guild Master. He said he had already searched. Three miners who would answer that profile are all off-station mining and have not had much communication. Certainly not enough to send this data, even in burst mode!"

Mackena frowned, "Oh, I also learned the word on the street is Allison has called her battleship and cruiser back from Earth Asteroid duty. She sent three destroyers to cover the earth-side duties. If the rumor is true, that will give her two big ships and three destroyers here. She looks to have mobilized her forces. You may be right about being the number one or two target."

Isaac began to type a query on his computer, stopped, and asked, "We will need to confirm this. I might end up running Mantle Works when this is done. What do we need for additional defenses?"

Mackena had reengaged the blueprint files and said, "These alexandrite lasers are a start—twice the range of Gaussian lasers. The beam is tighter with less Fresnel effect. That gives it more kick at more distance." Mackena paused and took a moment to zoom in on the image on her wrist computer. She turned the sand-can image around for Isaac to see. "You should have seen these changes even in the original design." A series of yellow circles appeared. "I missed them. Some of these are so simple – genius. Hey, think we can modify the sand cans into a two-stage offensive weapon with an EM pulse followed by the shotgun burst!" She moved on to another design.

Isaac was jotting down a note, intrigued by Mackena's weapon concept. Naturally, he intended to propose its inclusion in the trade agreement they had established. Historically, only Mantle Works had introduced new weaponry in the past six centuries. Although other firms had emerged, they invariably vanished under mysterious circumstances. A handful had even dared to challenge Mantle Works legally, but such efforts were short-lived. Isaac recalled the CEO who was purportedly driven to suicide by his own rocket—a narrative Isaac saw through. In the universe he was shaping, Isaac felt a profound sense of security.

Mackena took a deep breath and slowly released it, still focused on the design page. "Oh, this shield is a must! Wow, look at these new generators with the force shield. That is the most efficient I have ever seen. We can take on Allison's fleet defensively between this station and Trex. Can I start on these tonight?"

Isaac stated firmly, "All ready done. I thought you might be interested, so I cleared the printers for priority printing. Your team is in the printing area beginning to assemble. Said they were bored shopping for Ray Gunn, I believe."

"Good. I am thinking about asking the miners if they want a contract for the clean-up and salvage rights."

"That's bold," Isaac said, taking another chunk out of the pickle. "But I don't want all the violence. I think Allison's nature will draw her here, but after she has visited Cirrus Minor Station. All that firepower is a little much."

"You're a bit late with a peace offering, aren't you? I'd say she left the Cirrus Minor Station quite embarrassed, yet she managed to spare only the gunboats. Now, she's returned with her battleship and cruiser to resolve that issue. She assumes they're sitting ducks because they'd need to purchase new equipment directly from her." Mackena then redirected her attention to her shopping bag. "Such a shame. She played a high-stakes game and found herself outmaneuvered, unable to intimidate her way through. And now, one of her board guild members is stuck under quarantine. Little does she know, the station is now bristling with armaments." She glanced back at Isaac. "She will blunder into a hornet's nest there. You know the expression, Isaac?"

"What Expression?" Isaac responded half-heartedly.

"If you pursue the eradication of a hornet's nest, remember to make sure to take out Queen, or you have just pissed off the world's worst witch! Now, what did you sell the vug for? You never did tell me."

"The Chicago museum bought it for four hundred and fifty million, giving us a good profit. "

"I would say that was an excellent deal. Now you can take me to that expensive restaurant after I put these away." Mackena put the last of the porcelain birdhouses away.

Isaac asked, "Do you have a cup that says that too?"

"No, but I will suggest that," Mackena said, heading toward their bedroom to change.

Isaac finished the pickle with a satisfying crunch and prepared himself for dinner.

Chapter 21
Cirrus Minor: We Are Not Alone – Asteroid Claim – Cavern

Mining tractor designs changed over the years, but the primary purpose remained unchanged. The vehicles were designed for drilling lifting, mining tools platforms, and bulldozers. A safe, easy-to-use vehicle was used to extract and find ore in various ways.

The four youths and the dog entered the cave on two mining tractors, headlights bright in the dark. The shadows created a stark contrast against the rough cave walls. As they got to the narrower part of the cave, they noticed remnants of a torn air barrier. Jemma pointed to some empty cans and the spot where they found a body, and they all stopped. They quickly came up with a plan to turn off the Draconia plasma connection. It was unusual to find as many Draconite crystals as they did, counting up to 144, or another way to say it, 12 squared. This coincidence puzzled both Jemma and Tanya.

They discussed and decided to remove thirty-six crystals to weaken the plasma system's charge. The full-strength charge was lethal. When John asked why thirty-six, they explained that reducing the crystals would likely cut the system's power by at least a quarter. Since there were 144 crystals, a perfect square of twelve, they reasoned the best way to lower the power without risking the system entirely was to remove a perfect square of crystals—six squared, which equals thirty-six. They worried that taking too few might disrupt the circuit entirely and cause dangerous outcomes. To avoid uniform weakening, they planned to remove the crystals randomly, varying the charge strength of the Draconia batteries.

Two waited in the tractors while two collectors targeted the light point sources. John walked up to the first target. Unpacking a small container, he searched for the sparkle of Draconia but failed to see it immediately. "Tanya, I am at target number one. No obvious target … wait, Oh my, this is more interesting than it looks." John bent down and retrieved a small cloth-wrapped dark grey object from a small porcelain bowl lying on the ground. As he unwrapped it, the entire group could see the sparkle of Draconia from John's helmet lights. "Damn, that is bright. Oh, sorry, yes, but look for a little bowl."

Tony, watching everything from the tractor, offered, "We can probably retrieve the bowls with the Draconia wrapped in them and speed up the process. We have enough cold storage ready."

"I suggest we search the cave for any remaining items once we disarm this. They might provide us some hint as to who this person is." Tanya suggested.

The operation proceeded without a hitch. The miners managed to collect thirty-five additional Draconia crystals without needing to use a shovel. Meanwhile, Jemma kept watch, diligently monitoring the cavern for any unexpected surges of energy similar to what they had experienced the night before.

When they completed collecting the Draconia, they moved it half at a time to the axillary storage, isolated from the cavern and the eclogite. They returned.

John spoke somberly. "Now for the tough job. We have found someone in this cave. We do not know anything about this individual other than their different looks, and they amassed a tremendous amount of Draconite. If this were holy ground, we would see more ornate decorations. I think this person was stranded here. They deserve better. I may be desecrating a grave, but I believe my heart is in the right direction." He looked around. "I will put the man into the body bag for transportation home. I never envisioned we would ever need it. Does anyone want to help?" John offered.

Sue spoke, "I will. I seem to be familiar with death. I bet someone misses him. I will help."

The group split up. Tanya, Tony, and Jemma moved ahead, with Jemma in the lead. Using her sensors, she mapped out the cave and pinpointed the locations of man-made structures and devices. They found at the cave's end a makeshift bed and a small area resembling a kitchen, cobbled together from wooden crates that used to contain food supplies. Among these items was an object resembling a stove. The writing on the crates was in a language unfamiliar to all of them, including Jemma, who prided herself on knowing every human language. The kitchen setup bore resemblances to what they were accustomed to, yet it also had notable differences, hinting at the unique adaptations of its creator.

John spoke to them on the radio. "Our friend here is unusual. He has four fingers, an opposable thumb, two arms, two legs, and a tail."

"You're joking," Tony shot back.

Sue verified John's comment. "Yes, a tail. He has small equipment in his pockets we cannot identify. The suit he was wearing was very well-worn. We got him into a body bag, but it detached at the shoulder when we tried to straighten his arm. It was horrible, and we recommend that you not look."

Tony jokingly accused them, "Yeah, you two must feel like serial killers taking a prize?"

"Funny, Tony. Sleep with one eye open tonight," John replied.

"Better than that, Tony, sleep here," Tanya laughed.

"Okay, we have put him on your tractor, Tony, and will join you."

"I have one friend out here, at least."

Jemma then made an interesting note. "I have discovered a map, journal, and some old-style photos. I could use some help here. I don't want to convert to digital paws here."

"On my way, Jemma." Tanya moved to where Jemma was sitting.

"Sure enough, these are photos, and they all have tails. I cannot read the diary."

Tony then chimed in. I think we found the man's treasure. He held up a large metal box that he had opened. Packed in slots in the box were larger Draconia crystals. There were at least ten in the box. Tony bent low again and pulled out another box filled with diamonds loosely laid in the box. "Sweet!!!!! And three more boxes in here. So, was the man a miner?"

"We have not seen any cutting tools."

Jemma chipped in, "I may have located residues of chemical explosives. I recommend we avoid this area. There may be a danger if we find a stack of unused explosives. I have not sensed a bunker, however."

"I have a spacesuit with a tail slot. Jemma, it looks like you have a friend." Tanya alerted the group. "This suit has what looks like a map. I can recognize the cave and several features. I notice the landing spot. I think this may be the ticket to understanding some of this gentleman's language".

"Great, find we be there in a click," John said

Tony replied, huffing and puffing as he carried the metallic box to the tractor. "The first box is heavy!"

John arrived at the GPS mark he had made before Tony started to the sled. "Keep it up, Tony, Sue. Can you find Tanya? I will start carrying the other cases to the tractor."

"Sure, John, on my way," Tony replied.

The two split. John pulled out two more cases and started to the tractor. He passed Tony returning from depositing his load.

Meanwhile, Sue met Tanya. Together, they folded the galactic citizen's suit so it would join its owner. Jemma was now by the tractors investigating the ground in front of the tractors.

"There looks to be an imprint here of a vehicle landing pad. These are old marks. I detect ten distinct landing feet impressions. This vehicle had a low clearance vehicle height. If it could fit in here, it had less height than Adrian's sled. Maybe the Keyhole connotation is a good description. Did keyhole mean camp?"

"Jemma, that had to be a primary vehicle that brought supplies. Do you think this could be a miner's base because of how our friend was camped out? I am betting there are several mining camps around here."

Sue added, "Possibly, but why have we not found evidence of another society before today? This is an alien?"

"They may have been only after one thing like Draconia and diamonds." Tanya prognosed.

John questioned. "Did they solve our puzzle?"

"Possibly?" Jemma quietly answered.

"They did not appear to have an advanced technology."

Tony immediately countered. "If they had no technology, how did they get here? Why did this individual build a plasma array? That pattern was complex and natural. We have known about Draconia for a while and never dreamed it could be weaponized. This was not a primitive society."

"Like Us," Tanya finished.

Sue interjected, "If you are correct, we don't know their technology. We know they were here long before us; it looks like they left one behind. Worst of all, we do not know if we have disturbed this individual's dedicated resting place."

"Why would a society find and discover and mine a rare substance and leave it?" Tony reinforced.

John asked, changing the subject. "Tanya, did you get the books in the shelter?"

"Of course, we will learn something," Tanya jabbed.

Jemma chimed in," I have not seen any energy spikes. The background levels are at zero. We switched it off; thus, it is safe to collect the other draconite crystals."

John quickly reminded the group. "Again, note the position and orientation of the cup holding the crystals."

The youth also planned to collect all the Draconite crystals if they successfully defused the plasma generator. They moved, dropped back to the cavern floor, and pursued the other one hundred and eight crystals. They collected these in the same manner as the previous batch.

Once they arrived, they stored the body and treasure containers in the utility sled. John stored the Draconia first, knowing he would go into crystal thrall if he opened it. Tony and John continued to load the other four boxes in a unique, sealed container on board the storage sled. Tanya had thoughtfully declared a quarantine of the artifacts, which was part of the first contact process. They must keep these items sterile until a lab clears them. This included the Draconia and the diamonds because they had been stored in the alien miner's atmospheric enclosure.

Jemma carefully scanned the artifacts, with a special focus on any inscribed or written pages. The object Tanya found was notably fascinating. They made all these scans accessible to the public, with the map of the local quadrant uploaded first. Sue began the task of matching distant celestial bodies, such as significant stars, with their positions in relation to the Cirrus System. The correlation between the objects closer was impressive, reaching a ninety percent match. However, aligning objects in the near field of the Asteroid Belt turned out to be more complicated, prompting the four youths to carefully sift through the pages.

John overlaid the map to the asteroid's surface with the new data Jemma had recovered in the cavern. Sue joined him so they could plot their mining activities the next day. John and Sue determined they would start near the sled and work a square patch network to cover the asteroid. The starting point would allow them to survey eleven of the anomalies Turk and Adrian had discovered. Tieing into those anomalies would help them guarantee a well-calibrated, repeatable survey.

"Jemma, do you have any update on decoding the language," John asked

"I have some minor connections established but nothing that has cracked the code."

Tanya then added. "I was surprised our mining alien did not have any computer capability. They certainly did not leave any behind."

Tony asked, "We have evidence of a real alien race and a significant stash of Draconia and diamonds thanks to their efforts. Is it ethical for us to claim the minerals?"

"Good question," Tanya stated

John stood, "Well, there is not much more to do here tonight. I am going to turn it in. By the way, the alien miner was after the same thing we are - discovering unique minerals. I would want them to help a family instead of sitting in a box with a dead body. I guess that is my answer to Tanya's ethics question. We have commitments to fulfill for our parents, and we lost a day today from our schedule. It was well worth the loss, but we still have a job to do. Good night, my brothers and sisters." John turned and went to his cubicle.

Tanya had been tapping away at the air screen attached to her wrist computer. "I've drafted a message to send back home. Sue, can I get your thoughts on this? I'm aiming for a bit of mystery, so I need your take. It goes like this: Day 1: Our group got sidetracked by signs of an ancient mining operation at the cavern's far end. For safety, we've quarantined the area following an attempt to trace the camp's original settlers. We stumbled upon deteriorating explosives in a bunker, but rest assured, all's well. Our haul includes several unique mineral specimens and organic carbon fragments. A mummified miner was found, victim to a failed temporary shelter. We've secured and are transporting the remains for respectful burial, following strict contamination protocols for both the body and collected materials. The miner's faith or burial customs remain unknown. We'd appreciate your input on this. It seems best to bring the remains back to the station, stored in the auxiliary sled's quarantine compartment. Dad, this miner might well date back to before the early exploration of the Cirrus system."

"Hmm," Sue paused. "That is well written and does not give a bunch away. The last line will concern your mother, but she will figure it out. It might take her a while. Big day today. I am going to turn it in after writing my journal. Thank You for sharing that, Tanya."

"Night," Sue turned to Tony. "Are you ready to drill the first anomaly in the morning?"

"I attached the drill to the tractor after we stored our discoveries this evening. I am going to turn it in, too." He got up and wandered to his quarters.

Tanya followed the rest of the group, leaving Jemma curled up on the floor while processing the scans she had made. The map symbols, 3-D topography, and the survey across the asteroid scheduled to be shot by John and Sue were overlain. She sent that image to the wrist computers. It would give the team some additional things to think about. They would also be able to update anomalies instantly. Patterns would show up. It was a good exploration tool. The entire database would be available for any information chosen to store.

The halo display now displayed Jemma's manipulations with her teammates entering their sleep mode. Jemma scheduled her downtime with a reminder. She could get lost in this type of puzzle data. She had learned that rest was good for her ability to process the next day. She found that four hours of passive time was good. She began to search for meaning in the dead alien's journal. She started by comparing words on the map to the journal. She would look for a match in the journal. Then, she mapped repeated words in the journal. She was making statistical maps of the word distribution in the journal. She also mapped what appeared to be sentence structure. She mapped forward and backward, not knowing which direction the alien miner had recorded the journal. This was turning out to be a challenging problem.

Chapter 22
Cirrus Major - Controllers - Mantle Works
Headquarters -CEO office

The industry is known for its rather callous attitude toward employees. A piece of advice passed on from generation to generation is essential to understand. "You may work for a company, but you are not required to love them" (PAG 1985). Mantle Works had created its hellish purgatory for criminals on an Island. The prisoners called it Blood Island because everyone bled. The jobs the humans performed were dangerous and frequently exposed them to high doses of radiation. Weapons manufacturing was hazardous, and the enslaved people on the island experienced a short, miserable life.

Allison was in a great mood. Chadoom had reported successfully handling a problem at the Cirrus Minor Station. Under a false identity as a new miner, he had boarded a freighter and infiltrated the station. Chadoom made his way to the apartment of a Mining Guild board member who was self-quarantining. He noted the station's security was lacking. A security guard was found incapacitated, tied up in a classroom, dressed only in his underwear with 'unprofessional' written across his chest. Chadoom then assumed the role of the guard. After waiting five minutes, he accessed the apartment. The guild member was found with a wire around his neck, balanced on a block of dry ice. A note, seemingly written by the guild member, was left on the desk. Chadoom increased the apartment's temperature before leaving to complete the guard's shift. He exited the station on a freighter bound for Cirrus Major, ready to lead the task force

Allison was organizing to seize control of the space station and manage Isaacs's little hole in the wall.

Upon arriving at Cirrus Major, Chadoom made his way to the complex, where Allison required his assistance for another task. During his visit, he discreetly left several specialized surveillance bugs in the station to gather intelligence. However, he never received any reports back from these bugs, which infuriated him. The cause of this failure was one of Jemma's clever creations, known as Tanya's Cats. These were hunter-killer robotic bugs equipped with a powerful electromagnetic pulse (EMP) designed specifically to neutralize such surveillance devices.

"Thank you. I would like to see the controllers in fifteen minutes, please." She asked her assistant, "Let Chadoom in immediately when he arrives in five minutes. Would you mind freezing the controller's expense accounts? They will not need them as of today."

"Yes, mam" her assistant repied.

Four minutes later, Chadoom strolled in with a smile on his face. "Another successful mission."

"Yes, I read your message. Clever, you have the gall to finish that guard shift." She responded and continued, "Today, we need to help the controllers understand what their new job will be on the Island."

Chadoom smiled and continued his report, "I have taken care of the families as usual. They are being escorted to the spaceport as we talk. Soon, they will be off-planet and headed to Earth. One of the families, unfortunately, had a fire in their apartment when they refused to pack. The fire brigade could not get to the apartment in time. It is too bad to see such injustice create a fire. I am sad to say." Chadoom stopped and looked Allison in the eyes.

"How far can I go with the controllers?" Chadoom asked. He knew the answer to the question but asked out of politeness.

"I don't care. I need them to understand they are new citizens to the island." Allison said.

Chadoom thought briefly, "OK, I need about ten minutes to arrange some special uniforms."

"Fine, I have a little time to fill with them."

"Mam, the controllers are here." The receptionist announced to Allison through her internal ear jacks.

"Please send them in. Judi, they will be leaving via my private elevator." Allison sweetly replied. She smiled, knowing the truth behind her receptionist's belief that she was a widower. She also believed Allison was her savior because she had given her a job when her husband passed away unexpectedly. He had been in a fiery crash. Allison smiled, knowing her husband was a productive member of the Island. Chadoom had caught her husband trying to help a group of needy schoolchildren by providing them with a few pencils from the company. Chadoom had a problem: he had not asked permission to supply twenty pencils to the children. The man was given a choice. Allison understood and told him his wife would be cared for once he agreed to go to the island.

The five controllers entered the room and sat at Allison's round table in her office. They were seated stiffly, aware of something in the air.

"Gentlemen, Thank you for joining me today. I am sure you are aware Isaac Bauer has resigned. I have called you here to learn about Isaac's departure, which surprised me. I want to know what he was doing on the golf island and how he could move huge sums of cash around without you alerting me. My first question is, who do you work for?"

"We work for Mantle Works, of which you are the CEO," one of the gentlemen rose and said to Allison.

"Correct, I am the CEO, and you work for Mantle Works. Are you aware of how Isaac moved funds around the company?"

"Mam, Isaac was above the board in all dealings. He has been a great philanthropist and funded several projects from legitimate funds received from deals where obscene profits were made. We, controllers, worked with him daily and have not seen anything amiss."

"Did you know he may have built a large space-going ship under the island golf course under a non-profit that he headed up with his partner Mackena?"

"Mam, We don't know about a ship. The funds he used on your behalf are above-the-board non-profits with far-reaching influence in the asteroid belts of Earth and Cirrus Minor. It is not within the remit of our jobs at Mantle Works once the money has been properly accounted for. We have audited all the non-profits' books and found no discrepancies. The books are flawless, which reflects favorably on you."

Chadoom came out of the elevator with some bright orange suits and put them on the couch in Allison's office. He also had a cattle prod and a can of sweets. He went over to a man standing there, whispered to him, and they both walked back to the couch. After they talked for a bit, the man picked up an orange jumpsuit, took off his suit, and put on the orange one. He grabbed the cattle prod and went back to sit with the other people at the table. They all looked at him, wondering what was going on.

"My fellow controllers, we have a new mission. Please change as I have done. The rest will be explained to you as it has been me. We have worked together for several years, and I believe you should be able to trust me."

Three of the controllers stood up and changed into orange suits, following their colleague's lead. One stayed seated at the table. Then, one of the controllers who had changed got up and sat next to his coworker. He whispered something to the man, causing a sharp negative reaction. The controller in the orange suit said, "Forgive me, friend," before pushing the man's head down onto the table, which resulted in a broken nose. The room filled with the sound of the cattle prod as the controller in the orange jumpsuit pressed it against the man's stomach. Sparks flew from beneath the table, and the smell of burning flesh spread throughout the room as the man lost control of his bladder; a puddle formed under his chair.

"Please bring a coverall for Charlie. I am afraid he has ruined his suit and needs a change," Chadoom requested.

Charlie tried to strike his punisher but fell to the floor when his remaining strength abated. Allison had a satisfied look on her face at the drama unfolding before her. Chadoom grabbed the jumpsuit from one of the controllers and retrieved the cattle prod.

"Samuel, you realize this was at full strength. Charlie may need a little medical attention." Chadoom said, looking the controller in the eyes before turning and moving to the wall.

Chadoom pulled out a stretcher that was stashed in a hidden compartment behind the office wall. Samuel glanced at Allison, who had a medical kit ready on her desk in no time. He grabbed the kit, taking out healing foam and a quick-seal bandage. The other controllers took off the man's dirty clothes and gently fitted the legs of the jumpsuit onto him. Then, Samuel used a scalpel to carefully cut away the shirt that had fused to the man's skin, allowing them to remove it.

"The three of you load him on the stretcher and take the executive elevator to level minus two. A group of people will help you get to our next assignment and explain the mission. Gentlemen, we will be gone a while, and this is a top-secret Mantle Works mission." Samuel concluded, looking at Chadoom, who only nodded.

With their comrade on the stretcher, the three controllers went to the elevator, with Samuel lagging slightly behind.

"Samuel, please wait for me on sub-level one. We are going in a slightly different direction. Thank You," Chadoom requested.

Samuel and the controllers exited the office in the elevator.

"Chadoom, my unique friend, but how did you do it? I did not even have to terminate any of them to convince the others," Allison questioned.

"Samuel was easy. Remember the fire that killed that family this morning? I told him he was going to be arrested for murder. He would be made an example. I told him I would protect him and hire him to be my controller for the battleship. I only provided some clarity to him. He understood four things in my short conversation. First, his family was dead. Second, he would no longer be a controller for Mantle Works. Third, he had a short life expectancy if he disobeyed, and fourth, I was his savior. He just needed to achieve the goal I set for him to get the others to cooperate."

Allison said, "I was shocked that you allowed him to have a cattle prod, which he used extremely effectively, I might add."

"I ensured it was at full strength in case he needed it. And I learned something from him. Impressive, he might have a lot of potential."

Allison looked at her computer, "The battleship MF just entered orbit, and the cruiser is two days behind. Chadoom, it is important to remember that. I do not want the Cirrus Minor Station destroyed. I want them to be terrified by the firepower we surround the station with."

Chadoom wiped his brow and said, "Interesting. While on the station, I heard some miners discussing working on a defense system, but I could not gain anything else regarding intelligence."

"I thought we did a clean sweep. Did your destroyer gun crews fail to do the job?" Allison enquired.

Chadoom stepped forward, "No, we did a clean sweep. I will pick up my new sidekick and move to the MF."

"Oh, clean the table, please. I hate blood hanging around." Allison said, looking disgusted when she saw the table.

She suddenly turned, standing within a breath of Chadoom, leveling her grey eyes at his, "We leave for the station four days after the cruiser arrives and it is resupplied."

"Good, I could do with a rest. I need to finish work on the twenty A.I. fighters. All twenty A.I. fighters were transferred to the MF today." Chadoom released a set of cleaning bots who cleaned the table.

"Thank you," Allison said. She exited her office.

Chadoom finished and exited via Allison's private lift. He collected his new companion in the orange jumper, and they lifted off the planet in a shuttle thirty minutes later.

Chapter 23
Cirrus Minor - Simulation of a Fractured Planet - Mining Sled 2

Mining surveys still require an individual to set a pattern and follow through with typically geophysical tools like measuring magnetics, radiation surveys, and ground-penetrating surveys. The ground penetration surveys, besides drilling, relied on the sonic waveform (also ultrahigh frequency radar) to process signal arrivals against the original signal shapes to understand the velocity (radar: attenuation of electromagnetics) changes of materials in the Earth. Humans have significant experience in large aquatic and land-based seismic 3-D programs conducted on the Earth for several hundred years. However, a seismic asteroid survey only needed a small army of tiny robots, a vibration source, a computer for data processing, and a miner to drive the equipment. The miners were looking for voids or out-of-sequence arrivals that would indicate the presence of a crystal vug.

John and Sue finished their survey of the asteroid, mapping its surface geology. To their surprise, they identified several kimberlite veinlets, each with its own set of anomalies. The map from the alien miner showed similar findings, though it lacked the seismic grid that added depth to John and Sue's mapping efforts. They pinpointed a vertical cavity, or vug, three meters below the surface within a five-meter-wide kimberlite vein, having successfully extracted a smaller vug a few days before. Meanwhile, Tony and Tanya were working on a diamond-rich kimberlite found in a large crystalline cavity. This cavity featured fragments of the surrounding rock, known as xenoliths, embedded within it. Additionally, the magma around these crystals showed signs of wear, indicating they had tumbled through the magma, giving them a rough texture.

The ore went through a CT scanner to find the big crystals; then, it was ground up in a mill. They used a simple grease table, a traditional method, to catch the diamonds and then ran the leftover ore through another process to get even more out. The big diamond pieces were cleaned up and sorted out from the waste using a machine that separated them by weight. Diamonds caught by the grease table were cleaned and sorted right away. This way, they sorted the big diamonds by size and quality. They found only a little bit of Draconia, but it was still like a dream come true for miners. The hologram showed two things: one side was the map of the asteroid, and the other side showed what the Kimberlite vein looked like below the surface, the one John and Sue were working on.

Jemma was sleeping like a dead bug upside down. In reality, she was working on putting the sedimentary basins and old continental crust back in planetary space in the model they had been working on. They had nearly completed the first phase of their mining season.

Tony entered from the airlock. "We have filled our ore storage areas on the sled. Tanya and I just started to dismantle and store the plant. The space you asked for is about the only room we have left.

"Good, Tony, we need those chemical cutter bots for this extraction. We'll do it in a similar way: Dad and Uncle Adrian used to cut their vug but without the plasma. We will drill a small hole to get a live view of the inside when we get it trimmed. I believe we have a small probe that Jemma designed last week for this mission. However, we will get a good picture of the contents between Jemma and a cat scanner." John observed.

Tony replied. "What we don't have we can still manufacture if we still have the remaining stock regarding Dad's chemical cutters."

Jemma rolled over and scratched her ear with her paw. "Good. If you need more, I have a small revision to make them smaller and more effective."

Tony was caught by surprise. "Jemma, you take this dog thing too far sometimes. You could have just told us that over the speaker. Getting up and scratching your ear, give me a break. You're a robot. Next, you be sniffing and licking your butt like a dog too."

Now looking to solidify her dog act, Jemma walked over to Sue, put her head on her lap, and whined. Sue naturally reached down and scratched Jemma's ears. Jemma gave Tony a wicked smile.

Tony just looked. "Holy dog hair, that is just going too far. You have her wrapped around her dewclaw. John, that dog robot of yours just went too far."

Sue, who was now laughing, came to Jemma's defense. "Tony, perhaps she just demonstrated that she is more intelligent than you are. She has demonstrated her brilliance at taking on another life form's characteristics, including evoking emotional responses."

Knowing the dog had conned him, Tony replied, "Wolf, wolf, I am off to the showers."

Tanya, removing her hardcover, spouted out, "I agree, Jemma, no butt licking or sniffing, please."

Jemma then surprised everyone. "The only butt I would sniff would be Chadoom's when I bite him there."

John retorted, "While fundamentally I agree with you. It would be breaking a couple of robotic and human laws. Better think that one through."

Sue spoke somberly. "If Chadoom did create the artificial intelligence fighters that were used to kill my parents, as we have conjectured, I would bite his neck while you do his butt, Jemma."

Tanya stepped fully into the conversation, knowing it needed a turn. "Have you smelled our suits? Time to clean them. I thought I had died this morning when I put my softshell on." Both John and Sue agreed.

"What about our next step in our mining summer," Tanya asked.

"Sue and I were talking about that today. I think we are ready to run a breakup simulation. We have amassed a lot of information. Jemma pulled more out from our alien friend's map and books. We now believe this race probably witnessed the breakup and possibly caused it. They may even have tagged key blocks like this. When things settled a little, they dropped their miners."

"I would like to do an immersion session tonight. Now we have finished planning for tomorrow's extraction."

"I'm in," walked Tony, now wearing his pajamas. "Sorry about my moment of rudeness a few minutes ago. Jemma, it would be best to do everything you could about acting like a dog. It is in your best interest. I thus volunteer my butt for sniffing or biting."

Jemma immediately said, "Thank you. That is still not happening, regardless of how clean it may be."

They all laughed.

Tony said in retort, "Humor now? When will it end? Time for some foil to stop the voices."

"Won't work, Tony. I tried some in my suit helmet. She just talked faster," John said.

Once fully dressed, Tony sat in the immersion chair as the others joined him. The immersion session started with a beautiful planet fully reconstructed. John's cartoonish project significantly enhanced what he had shown at his last school project.

Sue began, "We've reconstructed this using some significant data pieces, never before integrated at a planetary scale." She sliced through one of the landmasses, revealing a cross-section of a sedimentary basin. "This represents a typical passive ocean margin. The sediment stretches from the crust into the oceanic basaltic crust, connected to a surface spreading center. This center indicates a convective magmatic core, which is crucial for plate tectonics. These divergent margins generate magmas with magnetic signatures or stripes as they cool, locking in the magnetic orientation at the time. As spreading happens here, as we've detailed in our model, subduction occurs at the opposite convergent margin. Basins under these conditions often have crustal accretion prisms and are usually rich in volcanic extrusions."

Sue paused briefly to gather her thoughts before continuing, "Naturally, when the lighter crustal material collides, subduction is often the result. This process involves the down-warping and eventual melting of crustal material, bringing together carbon-rich fluids in the deep, hot mantle. Mantle convection, influenced by subduction, is essential for incorporating carbon and other lighter elements into the system. Deep within the mantle, convection can cause decompressive melting. This happens when the warmer mantle material rises too quickly, unable to efficiently transfer heat to the surrounding mantle it passes through. Subduction facilitates the recycling of carbonates, leading to the formation of carbonated peridotite. This melt, being less dense, moves towards areas of lower pressure, leading to further melting. As carbonated peridotite melts, it produces liquids with a composition similar to that of kimberlite magma."

Once again, Sue paused, taking a moment to gather her thoughts before resuming, "The ancient cratons, made up of granites and gneisses, form a broad, sagged region along with the underlying asthenosphere, referred to as mantle keels. The existence of these mantle keels beneath the continents is crucial for the formation of kimberlite. They serve as a mechanical barrier that impedes the flow of the convecting mantle fluids. As the hot mantle material slows down, volatile compounds start to separate out. When melting initiates, these volatile elements move between the harder silica particles, sifting through the grain structures like passing through a sieve. This process leads to accumulating a volatile-rich, silica-poor kimberlite in a reducing environment. The growth patterns in diamonds, showing layers of resorption followed by new growth over the resorbed surfaces, highlight this dynamic. Diamond creation in the lithospheric mantle involves supercritical fluids or melts altering the mantle rocks they traverse, a process known as metasomatism.

In gem-quality monocrystalline diamonds, zoning patterns provide strong evidence that diamonds grow from an aqueous fluid. These patterns are crucial for analyzing changes in the carbon and nitrogen isotopic compositions during the growth and crystallization of individual mineral inclusions. Diamonds crystallize when carbon is released from the fluid, either by reducing CO_2 or through the oxidation of carbon-hydrogen compounds."

After a brief pause to highlight her next point, Sue added, "Therefore, we can anticipate different diamond-forming reactions in Peridotitic versus Eclogitic host rocks. For instance, a typical reaction in peridotites might involve enstatite and magnesite reacting to produce olivine and diamond, given the presence of a fluid. On the other hand, the mineral composition of eclogite is distinct, leading to a scenario where dolomite and coesite react to form diopside and diamond with fluid involved. In both contexts, CO_2 is emitted into the fluid, but diamond formation is contingent on the oxidation state being sufficiently low to maintain diamond stability over CO_2. The required oxidation conditions for these reactions differ significantly between eclogite and peridotite. Fluids overly oxidized for diamond formation in peridotites could still be adequately reduced to allow diamond formation in eclogites. It's possible for diamond-absent fluids to migrate from peridotite to eclogite, initiating diamond crystallization. This mechanism might account for the frequent discovery of eclogite xenoliths containing diamonds within metasomatic veins."

"At pressures where the diamond is stable, the cratonic lithosphere will likely have sufficient reducing conditions for carbon to exist as a diamond."

Sue was interrupted by John, "Here is the real kicker." John continued, "Most diamonds are not formed in Kimberlites. They are much older, like one to three billion years old. When created, the Kimberlites simply sample diamonds stored in conditions where diamonds are stable, ranging from about one hundred and forty to one hundred and sixty kilometers. This is a narrow envelope.

Some diamonds form deeper and are more often associated with peridotites under particular conditions. As the kimberlites originate from the superheated ecologites, the diamonds are disaggregated from the ecologites mass. The vugs, which were unknown, are rare but had to have formed in the eclogite. The vugs were swept along with the Kimberlite fluids, much like a rock carried in a glacier but at a higher velocity. These vugs could not survive long in the fluidized system, meaning the Kimberlite veins were close to the source. This asteroid was in the mantle keel or was very close."

Sue took over, "OK, we have one example of an asteroid we can plot as being in a root zone and several others where diamonds have been mined in asteroids containing near-surface pipes. The reconstructed planet began to rotate. We have reconstructed known basaltic oceanic crust with similar timing magnetic lines and any associated passive margin basins based on age and petroleum system when we could find the information."

The globe turned, highlighting the spreading centers and passive margin basins. "We derived most of the spreading length based on Earth models and associated basin age. We then tagged the thrust belts of various ages. We focused on accretionary prisms. We highlighted these, and again, we tagged all those asteroids together based on age. Then, we identified the old cratons and tied them together by age. We pulled these into the model based on straight-line ray paths from the current position of the Asteroids. That gave us a bit of confusion until we started working on potential orbital arcs. It was not perfect, mainly due to post-impact collisions, but we could group the asteroids and bring them into the model based on a series of conical arcs. Similar to predicting the flight of a pellet from a shotgun blast."

Sue was excited as she continued, "That left a planet rotating on the screen with the planetary crust exposed except for a large round white patch on the planet. "The white patch is an impact site. Most models bring a rogue body in from the outer Ort belt beyond our sixth planet at a high velocity. We don't think that is the case. Instead, we bring a rogue body at a low level to the galactic axis near the orbit of Cirrus Minor. Cirrus Minor could have captured the planetary body under certain conditions if small enough. Our model demonstrates the slow rupturing of Cirrus Minor. The model showed the two bodies slowly disassembling due to significant gravitational pull. We think this other body was dominated by ice due to the amount of ice found in the asteroid belt. Eventually, with its substantial velocity, the more petite body contacted Cirrus Minor, causing the planet to fragment and disperse catastrophically. The rugged Nickel Iron Asteroids is an example of the last phase where the molten core was torn open and ejected."

Sue continued to change slides as she described events, "Meanwhile, a fine debris cloud formed at the collision site and was incorporated into a series of rings when things settled down. The asteroids had two significant movement vectors. The first was associated with the collision when they were propelled along conical straight lines, and the second as gravity realigned them into new orbits. Significant mixing occurred due to the rogue body plunging through the disaggregated Cirrus Minor. I will put it into motion without the bright lights and loud noises that John used in his early conceptualization of this model. Note we have highlighted potential mantle keels in the diamond preservation zone in yellow." Sue sped the model up with the yellow highlighted zones, dispersing rapidly into the asteroid field.

"We then mapped probable areas in the asteroid belt where we should find more of these keel-type zones. We eliminated anything that is known. This dropped several yellow targets. The residual are potential anomalies." They ran it several times. "Please note we have anomalies around this asteroid's neighbors. The mining database does have some data that does not completely fit. Some of it is bad data, and others are incomplete records. Still, numerous asteroids around us may have potential. We must prioritize the highest potential areas and prove this model is a solid exploration tool. Once we have some priorities, we must determine how to pick the winners to establish a series of claims."

Tony got excited about this. "I have some ideas about bots doing our exploration work."

Tanya interjected, "There's so much to explore on this map, and obviously, we can't tackle it all ourselves. But our family will make the time. I imagine that once we've shown this can work, we'll bring on miners to work on the sites while we scout out the rest of the map. If I'm right, we're going to leave a legacy that extends beyond just the four of us, including future partners. So, let's zero in on a few key areas. First, our immediate vicinity, and second, the route through the keyhole into the heart of this dense asteroid cluster." The keyhole was a slim passage clear of asteroids, but its challenge lay in being encircled by a thick ring of asteroids notorious for erratic movements. It earned the nickname 'sled killer' due to the wayward asteroids that would shoot through the passage, smashing into fragments on the other side, making it hard for radar to alert to these rogue bodies in time. It was a notoriously dangerous route.

"Hmm, The keyhole!" John commented, "Dad hates that place."

Jemma jumped up like she heard a shot. "Tanya, you are brilliant. You have just connected the dots, so to speak."

The hologram changed to the alien miner's map of the asteroid belt. The keyhole was highlighted, but it had a different configuration. Jemma, as excited as artificial intelligence can act

(Now defined as the AIE factor was a seven), "We don't know when this map was made, but we can correlate larger asteroids that have been fixed. Our model generally predicts this feature, but it is neither like the Alien Miner's map nor our own charts. That is due to rogue asteroid activity, which is a random event."

The map changed to Adrian's miner's map, which he was following when he jammed his sled into the cave opening on the asteroid they were sitting on.

Jemma, still excited, continued, "Adrian misread this map. He was correct that this asteroid was important but only as a navigation beacon to line up on the Keyhole. His primary objective is on the other side of the keyhole, an area we would mark as a priority."

"What about the dangers of the keyhole? We have a substantially valuable load stored in our sled, and it will be even more so by the end of tomorrow. Do you want to risk that?"

"Oh, Tanya, you only live once. What good is all this if we just stash it away and settle for a comfortable life? We're on the brink of something greater," Tony challenged. "Our journey began with two goals: to mine something and to explore. We've achieved the first, and now we have the means to pursue the second with unwavering focus."

Tanya retorted, "Tony, a chip of crazy Adrian did fall off the tree after all. I am only concerned about the Keyhole. Have you seen all of those notches painted on Dad's ship? Those are rogue asteroids he has blasted. Someone close to him was killed by one, and now he hunts them."

"I believe we have mitigated the risks with improvements to the standard mining sled," John commented.

Jemma smiled a toothy smile that caught the light just right, creating a glint of reflected light like someone in the movies, "Exactly, John, our mining sled is equipped with cutting-edge sensors and weapons unlike anything else out there. Even without missiles, our current arsenal is nearly on par with a cruiser. We have the capability to outdistance the main guns of a battleship with most of our armaments. We should be able to spot a rogue asteroid well before it becomes a problem. And if one does slip by, remember, we've got an AI on board trained for combat pilots. This sled isn't just fast for its size; it's faster than almost any craft humans have made, even when fully loaded. Plus, we could send out probes to navigate through dense debris, enhancing our early warning system if necessary."

Tanya and Sue were unaware of the modifications Jemma had made to the ship. They glanced at the dog, whose tail was wagging playfully. Sharing a knowing look, they simultaneously declared, "I'm in."

Tony turned to John, "Well?"

John caught up in his thoughts and snapped out of it, "I'm in, but we have to keep Dad out of the loop until we are through. Otherwise, he will bundle Crazy Adrian up and come hunting. Neither he nor Adrian have our capabilities. Jemma failed to mention the improved force field that will keep the debris away from us. Jemma, can we build a gunners station in case you get knocked out for redundancy."

"Yes, I have a design in mind. It is ready to be deployed in the immersion center."

"Ok, Sue, tomorrow I will get the plasma cannon mounted on a gravity sled with the chemical bots. We will retrieve the targeted vug if that is what it is."

"Ok"

"Tony and I will complete storing the mill and will tighten up the storage," Tanya indicated.

Tony could not resist. "Jemma will laze around and sleep like a dog all day."

They all laughed, knowing that is what she does when running the planetary simulator. After all, she was one of the most sophisticated computers in the galaxy and learning FAST.

Chapter 24
Cirrus Minor - Call to Arms – Trex control room
Asteroid 16

"Follow the money." Money was the cause of every skirmish man's participation since becoming a spacefaring race. Greed and control were the main forces at play. Sadly, something as simple as money drove people to commit horrible atrocities. Space warfare was designed to bully and force others to yield and give up. Power went to the company that was the biggest bully – Mantle Core. Mantle Core defined a new form of hostile takeover by using their fleet. Since they sold the weapons, they controlled the outcome. Anyone inventing a new technology or weapon was of interest to them.

"**M**ackena, we're about to start a war with a company heavily armed with a battleship, a cruiser, and three destroyers right here in this system. That's not even counting their additional destroyers under repair. This company, which has amassed a great deal of wealth, some of which I helped create, is no small adversary. To give you some perspective, the Earth Navy's entire fleet consists of just two battleships and three destroyers. Much like Mantle Works dominate the Wild West, the Asteroid Belt near Earth, they control over half of this territory with their battleship and cruiser, ruling it with an iron fist. Meanwhile, the Earth fleet is preoccupied with defending the other half of the Asteroid Belt and dealing with a rebellion on Mars, hampered by their older technology. Despite their discomfort with Mantle Works' armed forces in their territory, they can't afford a conflict against a fleet with Mantle Works' firepower."

Mackena smiled, understanding they shared a common concern, "Yes, Isaac, we are going to protect our interests. We are contractually linked to the mining guild. The local guild was pushing to join us. They have about a hundred miners with sleds they want to help in this conflict. I set them up with defense commander Hall Gunnerfield. He is going to set up a backdoor ambush. They are hollowing two iron-nickel asteroids to set 50 sleds in each. The ship's plasma cutters are being modified to the new configuration from the guild plans."

Isaac stopped momentarily, thinking about Mackenas' response, and finally cautiously stated, "You seem excited to get into a shooting war with ships that have really big teeth."

Mackena took a deep breath before replying, "I'm just curious to see how my baby will perform. It's almost as exciting as shopping for new Christmas decorations. Maybe I should get the chance to blow something up for once, especially with those guild modifications we acquired. Those alexandrite lasers are truly cutting-edge. Researching synthetic Alexandrite to create them wasn't easy; we had to dig through an old 1995 Honeywell corporation document to find the procedures. These lasers outmatch anything we've encountered before, delivering unprecedented power. Interestingly, Earth's Navy has a research group exploring similar technologies. Fortunately, they seem to have taken the wrong turn and are currently stuck."

Isaac smiled, interjecting, "Interesting. What's the potential value of supplying these technologies to the Earth Navy? We could potentially earn several hundred million credits. However, that would require establishing a new manufacturing unit and acquiring an additional satellite. It's a promising idea, but we need to stay focused on our current challenges."

"Isaac, you're a business dope. I have figured out a potential defense for the guild at the Cirrus Miner base. They have an untested defense and an untested ship that I have one hundred percent confidence in. But we have a weapons advantage. This means we bait them into a trap once Mantle Works establishes a clear threat to the station. Like the miners guild here, the miner's guild of the station wishes to help. They cannot evacuate the families from the station. They will go to the special shelters they are building. Their miners will be the door to the trap. Depending on how Allison's fleet deploys will determine how we use the guild here. They are our future, you know."

Isaac, a little confused, "OK, how do we handle the firepower?"

"What would you do? Mackena responded with a question.

Isaac responded promptly, having already simulated the scenario multiple times. "I'd initiate a missile run with my destroyers targeting the station's defensive positions, followed by deploying assault troops. This would position my Battleship for cover while the cruiser stands by for backup, anticipating that the guild miners won't yield easily this time. I suspect they might have a hunting party lurking in the asteroids, waiting for us to engage the station. Assuming they've reconstructed something since my last visit, the Cruiser will take a high or lateral orbit to remain agile."

Mackena agreed, nodding, "That sounds like what I would do. If the destroyer fails, I would hold the battleship's fighters as missile cover and backup. Missiles are easy to make, and I would expect the guild miners would attach a few to sleds and make a run."

"Correct, and that is what we are going to give Allison. Twenty-five sleds will launch 200 missiles at the battleship. After arriving, we will throw missiles off racks and distribute remote missile pods. I believe we will fire another 300 at the Battleship. That leaves our main ship stocks as a backup."

"Let's keep their sled squadron on standby for missile defense. While this cruiser boasts a solid shield and an array of anti-ship missiles, it's important to remember it's still just a cruiser. Battleships tend to loathe surprises, especially from cruisers they know little about. The miners' base will likely be preoccupied dealing with their destroyers, so they'll be in for a rude awakening when they realize they're up against a fully fortified station and a formidable ship."

"That's why we'll go silent as soon as we detect them. We'll maneuver behind the station until their destroyers are within an optimal range of the alexandrite lasers. Anticipating they'll follow up with assault shuttles after a missile barrage, we'll rely on our EM weapon to dismantle their missiles. Meanwhile, the station will keep its focus on the lead destroyer at all times while we launch our attack on the Battleship."

"What if they lead with the Battleship?" Isaac asked

"Simple. If they led the battleship, they intended to destroy the station. We destroy the Battleship with a close-range missile attack plus our weaponry. We will want to move fast, and the sleds and gunships attack gun turrets on the Battleship."

Isaac now understood Mackena had been very busy, "Sounds like you worked on that."

"We did this morning when you were eating pickles and making money. We had a remote meeting that station guild members joined. One of them joined us remotely and stood out as a strategic thinker. His name was Golden. He was running simulations and providing us with survival probabilities faster than I had ever seen. I think this is the same person who came up with the guild's weapon system. However, if you look for information on him, he is basically off the grid."

"If I had a golden laying goose, I would treat them with great love and hide them," Isaac said, hugging his wife.

"Is that not what I have done with you, dear?"

Chapter 25
Cirrus Minor - In the Keyhole - Mining Sled 2

The Cirrus Minor Asteroid field had a few absolute dead patches containing minor debris but lacked asteroids. The Keyhole was one of these features. It was an extended, cylindrical, narrow, open area of space surrounded by dense concentrations of asteroids. The Keyhole looked like a peaceful spot to put a sled in cruise control and nap. It is also known as the ship killer. Too many miners had died trying to cross this section of space. Rogue asteroids ranged through space. They are, by their nature, unpredictable, with paths that change as they play billard with other asteroids. A unique, long-standing gravity wave kept the route open, but the keyhole's edges were lined with dust from the collision of thousands of asteroids. Turk's best friend had been killed while trying to cut across the Keyhole.

The search through the initial group of anomalies yielded exciting results, including the discovery of three more asteroids composed of Eclogite. Among them, two asteroids featured kimberlite dikes, with one even boasting loose diamonds on its surface. Additionally, three additional claims were spotted, each designating the youth as twenty-five percent owners, with a crossover assigned between them. A new weapons station had been constructed, equipped with two full-immersion virtual seats where the youth had been diligently training on the weapons system. Aligning the original claim and the Keyhole, they executed a close flyby while Jemma showcased a 360-degree halo display. This display depicted the Keyhole the ship was traversing, along with the surrounding asteroids and dust layers, highlighting any rogue asteroids, none of which were colored red.

Twenty minutes later, the sled entered the keyhole's narrow portion. All had been quiet. They had tracked four rogue bodies, but none were a threat. Three of those rogue bodies self-destructed into some giant asteroid bodies that lined the keyhole. These appeared to be spectacular crashes, as portrayed by the sensor data. They understood what created the high volume of dust after seeing these.

The lights in the ship shifted red. "Battle stations, Battle stations; report readiness!" An eerie voice chided the astronauts through the speakers."

"Tony, pilot, good to go"

"Tanya, copilot ready."

"John sensor weapons station one ready and tracking rogue forward quadrant three o'clock galaxy plane plus thirty degrees. Fifteen thousand clicks out."

"Sue sensor weapons station two ready and tracking second rogue rear quadrant seven o'clock minus fourth seven degrees. Twenty thousand klicks out."

"What two rogues are on a collision path with us in three minutes? Unreal. Recommendations?" Tony asked.

Jemma piped in. "We have the firepower to deal with these, Tony. Straight and true. Let's let our ship work for us and not try to outrun these things. We eliminate the threat and are through the door in ten minutes."

"Affirmative." You could hear the concern in Tony's voice. Jemma had trained the crew to military standards during the prospecting time. She drove them crazy, drill after drill after they had agreed to the training. They all were trained in all the stations.

John, feeling most at ease in the sensors weapons arena, suggested, "I recommend we use the AlexLaser and the pulse plasma cannon to take out the first asteroid. That way, we can reserve the main plasma cannon and the underbelly cannon to handle the second asteroid. Sue, what are your thoughts?"

"Sounds good. We may need to go into a glide mode and rotate the ship down fifteen degrees to optimize the use of the two cannons. Tony, what do you think."

"Yeah. Tanya?" Tony agreed, asking his copilot if she concurred.

Tanya paused for a moment, contemplating the strategy. "Sue, I'll take charge of the rotation while Tony manages the glide. It's crucial that we ensure our forward shield is prepared to handle the debris from the first asteroid. After firing the initial shot, we'll have three seconds to adjust the sled off the zero glide plane, followed by another three-second window to take our second shot, and finally, four seconds to realign to the original position. It's a tight ten-second window, so we need to execute with precision."

"I am good for it," Sue replied, "I will set the shot up."

"That gives us roughly a ten-second window from the first shot on the forward asteroid. We only get a second shot with the AL-lasers if we miss that or fail to cut it to pieces. The pulse cannon will only be partially charged. We have no wiggle room."

Jemma opened a private channel for John, saying, "You have a 99% chance with your plan. I can always take flight control and do things you cannot. There would be potential injury and high g's that would be extremely uncomfortable to humans. I like your plan and think you and your group have exceeded my expectations. But then again, I don't have a large database of live experience to draw from".

"Thank you, Jemma. It is reassuring to know we have a hotshot backup. I now understand the consequence for screwing up is punishment by high g maneuvers."

Sue suggested, "I will load the magnetic cannon for an emergency shot if we miss. Tanya and Tony, you must rotate the ship 180 for the shot. It would also position our shield in optimum position."

Spence was quick with his reply, "We will get it done. Sue, you will need to make the call."

Jemma was intrigued by the fact that the humans were so good at war. She finally realized that she was born a warrior by a warrior race. These humans, however, were loving and peaceful and played by moral and ethical standards. Being a robot without rules could lead one to dark places quickly. She had been watching old Earth Science Fiction. She loved a series called "Star Wars." While she did not like the Dark Lord, she was attracted to the secretive style the movies made him. It was similar to a cartoon she read in the high school library under

John's name called Dr. Death – the plague doctor. She was afraid of becoming that evil Emporer. The dog shivered a little, knowing she would be safe if she stayed near her friends. She remembered a phrase "Choose the Right" Sue had used. CTR would be something she would etch on a dog tag.

The hologram in the room flashed when the first rogue asteroids broke free of the keyhole's debris field. Thirty seconds later, the asteroid below and behind the ship also penetrated the keyhole. It was massive, as expected. Combining the two asteroids on a collision course in the keyhole would have guaranteed the destruction of a standard sled. The modifications of the Ericsson sled two made this a different proposition. It was armed with weapons with offensive or defensive capabilities. In this case, they would aggressively remove the rogue bodies entering the keyhole with extreme prejudice, like John's and Tanya's father. His dad's currently configured sled would not have survived this encounter.

"Al-lasers firing in three," John announced. "Two one." Two purple beams shot out from the front of the sled. They hit the dead center of the asteroid as it spun, causing a deep groove to appear in the dense asteroid. The asteroid rotated three times on its fast-spinning axis before cleaving in half. The second plasma cannon typically stored in the haul had been prepositioned and then pulsed a series of plasma bolts. It spread the bolts across the two halves equally.

Tony excitedly announced, "Thrust in Neutral. Begin glide now."

Tanya added, "Rotating sled positive fifteen degrees rotating now."

The plasma bolts hit the first asteroid's two halves, immediately breaking into smaller rock chunks and dust streaming toward the sled. Three large chunks were a direct threat to the sled.

Jemma announced, "95% chance of needing Alexandrite laser shot to clean up larger asteroid debris."

Sue calmly announced," Firing primary pulse synced with belly plasma NOW!" Lights in the sled dimmed only very slightly. Two green beams shot out from the sled. The giant beam struck the asteroid high and on the right of the center. The smaller belly plasma pulse stuck the fast-moving asteroid just to the left of the center of the mass.

Tanya announced, "Rotating sled to the plane."

"Thrust active," Tony declared.

"The smaller asteroid sheared into small fragments and dust. It is no longer a threat." Sue happily announced.

"Good shot, Sue," Tony calmly stated.

Things were still moving fast. The debris from the destruction of the first asteroid was closing on the mining sled. Three large angular chunks were still on a collision course with the sled. Each asteroid fragment was cut straight through by the power of the newly configured weapons. The giant asteroid fragments were all spinning along an axis perpendicular to the sled. Since plasma was a beam, the first set of shots was designed with a cutting motion. The direction of that cut was the direction in which the rock was spinning. Destiny was in the Little Minings Sled future. John was going to play some billiards with the three rocks. He would take the two shots, targeting the two outside fragments just off-center. These shots would ricochet the extraneous chunks toward the center fragment. As a result of the collision, the remainder of the two outside pieces would hopefully change direction away from the current target.

"Alex-laser shot two lined up. Shooting two targets in three two one."

The purple beams shot out again, striking two of the outside chunks. These disintegrated, sandwiching the middle fragment as planned. A couple of larger fragments escaped with low velocity. The debris field was rapidly becoming a dust cloud expanding from the center. The tiny mining sled with its cargo underneath quickly flew away from the cloud.

"Holy laser dual, that was impressive, John," Tony said, gripping the two control sticks with white hands.

"Heavy debris ahead! Boost shields to the front, and check your harness. This will get bumpy," Tanya announced. "The Keyhole exit has accumulated a lot of minute particles like dust, except it is thick."

"Need a path cut? Ready now." John quickly plotted a pattern to clear the debris.

"Go, now, John, this stuff is thick – extremely thick." Tanya returned.

"Starting now." John touched the switch in his mind.

Purple beams began to clear the path, filled like a dusty corner in a house with a long-haired dog. The front shield flared bright white when the residual debris hit the screen. The little sled severely vibrated as the shield flashed. The sled was through the keyhole, and the trap was disarmed. Lights in the ship shifted from red back to normal.

"Scanning the target area," John announced, needing to stretch out to the bathroom.

Sue was already out of her gear and stretching after the horrific shaking the little sled had taken.

Tanya announced, "Taking pilot controls, Tony is taking a stretch. Turning to the area of investigation. We will be there in 10 minutes."

Tony walked through to the galley, grabbing a bottle of water. John got out of the immersion rig and walked to the bathroom.

Jemma initiated a higher-level scan after picking up a reflective light anomaly in the target area.

John returned. Sue moved to the crew area as she was officially off duty. Tanya joined her on auto-pilot with the ship (which meant Jemma was piloting and loving it).

John picked up a similar anomaly that Jemma had seen. "Hey guys, we have a reflective light anomaly in the strike area. I want to launch a probe ahead of us to investigate."

"Sure," the young ladies turned to the halo screen for a look.

"No damage from keyhole crossing," Jemma announced

The group focused on developing a plan for the second target area. The probes provided additional data so a good map could be compiled. The early launched probe recorded another three reflected light flashes as it neared the target area.

"No registered claims for 150,000,000 kilometers. There are only two registered just outside our area of interest. Those claims are iron-nickel. This area has rarely been explored. It is accessed by either the Keyhole or the Jungle. The Keyhole is easy if no rogues are present. The Jungle route takes skill because the route is tortuous, snaking around several dense clusters of asteroids. The asteroids are stable, but the route requires concentration to navigate," Tony concluded.

Jemma highlighted the Jungle with lines of navigation that were complex and angular. "It is a shorter route to return to the mining station or Asteroid 16, our new home. Navigation is fairly straightforward, but common sensors are limited in the jungle. It is a safer route but requires a skilled pilot, for which we have five on board." The dog on the couch wagged her tail as she rocked the ship side to side.

"Funny," Chipped Tony. "Damn D…"

John interrupted. "Interesting probe one has initiated a surface landing. Jemma switched to the photos the probe had taken. They are interesting. The hologram displayed the first picture, depicting a dark asteroid full of veins of lighter material. The veins radiated from a mass at forty-five degrees to a central intrusion. The black mass is eclogite; the lighter mass may be kimberlite or lamproite. The probe has instructions for sampling both substances. We should investigate this as it has significant potential!"

Sue was excited by this news. "This looks like our planetary model is predictive! The destruction of the Cirrus Minor Planet was either a tragedy or cold-blooded destruction of a healthy planet."

Tony was always ready to throw water on a fire. "But Sue, we need to confirm the information."

John, whose eyes had just doubled in size, interrupted. "Maybe we have. Look at this new photo from the probe."

The hologram changed to a block of eclogite in the flat foreground, transitioning into a lighter-colored mass ending in a three to five-meter cliff face with a crystal mass. The crystal forms were hexagonal, not tetrahedral or octagonal, and reflected starlight as the asteroid rotated on its axis.

Tanya broke the silence. "That looks like Draconia. That is a lot of Draconia if that is what it is."

"We will know in a few minutes. I have instructed the probe to sample the cliff face". John replied.

Tony then interjected, "I have launched the probe pattern we designed. Ten launched. Let's plan on claiming this asteroid while we are probing. What else can that be other than Draconia folks."

Sue was watching the probe's progress. "Close-up photos of the cliff face leave little doubt of the crystal form. Sampling probe X-ray diffraction of the lighter colored material in ninety-four percent confirmed Kimberlite. The Eclogite is the dark mass confirmed. Let's launch the claim buoys."

Tony then said. "We are down to six remaining sets of claim buoys. This is the best evidence I have seen in our prospecting. I'll set the coordinates, and we have plenty of time to relax while we set the claim markers. Is anyone ready to land near that little cliff face in the shadow?"

The other three nodded in affirmation. They set the first set of claim buoys and proceeded to land. Tony and Tanya landed the sled on the flat with their repeatedly practiced expertise. The procedure was first to land the underbelly sled, then the mining sled. That went without a hitch.

The group, including Jemma, dawned on their soft and hard shells to explore the cliff face. John ensured everyone double-checked their connections, concerned that their excitement might lead to an accident. His dad was coming out in him. They checked out except for a small hole in one of John's oxygen backup hoses. Before they exited, they switched out the defective component and stopped to take a moment to review their action plan to ensure they did not go off like cowboys. It was decided that Tanya and Sue would form one team and John and Tony the other. Jemma would go with John and Tony to explore a 500 m radius around the cliff face while the other group sampled the face. Then, they exited the sled.

"Wow, I think we have a working model to predict Mantle Keels," Sue exclaimed. She had reason to be excited. It was the thrill of discovery that was driving the group. They were beyond wealthy with what they had found. They had collaborated in a way the Universities were still thinking about. They had put the astrophysical model together and used it successfully as a predictive tool. What else could it be used for? Now, they would reap a few rewards!

The face was comprised of Draconia crystals clustered in masses of diamonds. They were the inner mineralization zone or were developed originally inside the cavity. The outer zone consisted of massive diamond clusters. The Draconia under their suit lights was dazzling and mesmerized the group. They stood for five minutes without a word.

Jemma broke the silence, "John, I am receiving an all-out mining sled recall in the belt. The Mining Guild is issuing it. This is an emergency call back."

"Can you confirm the authenticity of the guild order?" Guild orders of this magnitude were unheard of. A general call back could only mean an impending disaster or a call to arms.

"Signal confirmed. It is for all mining sled guild members from both locations." Jemma replied.

"Okay, I will tell the others. Can you finish recalling all the probes and prepare the ship for departure?"

"Yes, I will head to the ship." John saw Jemma turn and head back to the ship.

"Sue, Tanya, and Tony, The guild has just issued an all-sled general recall order. We must pick up and get moving to the guild-defined check-in point as soon as possible."

Tony responded, "Good, we have time to strip this face and retire."

Tanya jabbed him in the side. "No, Tony, we need to figure out what is here and how to optimize the extraction. We have enough time to collect a sample and concisely survey this local area."

Tony, in his youthful enthusiasm, understood Tanya's well-thought-out comments. "Ok, John and I will sample the top of the cliff and drop a poly-canvas from the top to cover this face."

John said, "I am good to go. Jemma is recalling the probes and preparing the ship."

"If she takes off, we become mummies like our friend, only to be discovered thousands of years from now," Tony said, trying to lighten the mood.

"Not funny, Tony," Tanya immediately reprimanded Tony.

Jemma jumped in. "Not a bad idea, but you have grown on me, Tony. I like you around. Anyway, I owe John and his family a huge debt."

Tony shot back immediately. "Not only is she a robot, but she has a wicked uppercut, John! Who programmed her? You?"

Sue surprised the group. "Tanya, look at this red crystal shaped like a pyritohedron. Have you ever seen anything like it in your research?"

"No, what is it?"

"Don't know. I am going to sample this area of the cliff base". She pulled out a spray can and sprayed a one-meter square thin line.

"I will prep the core drill so we can core through to the host rock."

"Do we have time?'

"Yeah, John and Tony cannot move that fast."

John and Tony grabbed some stored poly-canvas and pitons to secure the canvas. Tony grabbed a climbing cable, attached it to the winch, and dragged the lightweight cable to the top of the small cliff. He drove two pitons and connected the line with two carabiners. He put on a climber's suspension diaper and gently repelled down one and a half meters, where he assumed a sitting position. He painted a thin square with bright yellow fluorescent paint and photographed the meter square sampling face like Sue. He then dropped a small gravity pod with a meter extension at the line's base. This was a sampling ledge that would help him collect all the fragments he might knock loose.

Meanwhile, John drove a series of pitons with clips across the top of the outcrop and then tagged three at the base.

Tanya cut a half-meter core terminating in a Kimberlite matrix, which she then labeled and photographed. She put an arrow on it, indicating which direction was up. Sue removed the Draconia crystals, a few associated diamond crystals, and ten pieces of a transparent red mineral. She put the red mineral in a small container.

Tony finished his sampling but failed to see any of the red minerals. He had discovered some attractive pink diamond crystals under the Draconia. He had John grab his packed specimens while he captured the material that had fallen onto the small gravity probe. He climbed out and assisted John in attaching and dropping the tarp. The tarp was to prevent the reflections that they had initially seen. John and Tony attached the tarp to the base and picked up the equipment. Tanya and Sue had headed to the ship with their new treasure. They, too, carried their one-meter sampling container to the vessel, where they would spend hours sorting through it after it had been weighed and cataloged.

John removed his hard shell and asked, "Jemma, are we nearly ready for departure?"

Jemma is like a good crewman. "Aye, Captain, drives are warm, the storage sled is reconnected, and we are ready for departure."

Then, she had her dog counterpart walk into the room with a guilty dog look. This is when they will not look you in the eye and slink around after eating two pounds of leftover ham from the kitchen counter. When caught, they roll over on their backs into the dead bug position, hoping for a reprieve. She did this so well that all four of the youth were shocked. They all just froze in their place.

Tony broke the ice. "Did you poop somewhere or chew up someone's squeaky toy?"

Jemma said, "Oh no, I have been a good dog, maybe too good. 'This recall is about Allison Jiggs driving five ships at the station. You don't know it, but she has called in her battleship and cruiser and is now on her way to the Cirrus Minor Station. The other day, I participated in a call with the guild and Isaac about defending the station. I did not anticipate a general recall order. This had gotten serious. I participated as an unidentified guild member named Golden. Isaac and Mackena probably figured out I was the same one that redesigned the station's defensive weaponry."

"How did you communicate at this distance?" John asked

"I have a few tricks up my sleeves that have not been publicly released. We can call anywhere in the Cirrus Minor cluster now." Jemma frankly replied

"Gasp, she now has "reach out and touch somebody" communications. So we can call the folks if we wish." Tony jumped in, looking to lighten the mood again.

"Technically, yes. But not in high-density asteroid fields. We are in such a field." Jemma indicated. She continued, "Here is the issue that has me concerned. The defenses of the mining station are strong, but Allison is bringing a Battleship, Cruiser, and three destroyers. She will attack the station with the three destroyers. She will use the Battleship to cover any station issues and the mining sleds everyone knows will attack. Isaac has a unique design for a ship that is effectively more powerful than a Cruiser. He will take on the Battleship with the help of the miners. That leaves the Cruiser as the wildcard without a plan. The group suggested additional recalled miners will deal with the Cruiser once they know what it will do. The problem is the Cruiser could change the outcome in several ways. It needs to be taken out. We have the firepower but need an edge to close, so we are not fighting missiles. This mining sled is a serious warship, too. How do we get close?"

Tony was reserved, "Why do we have to do this? We have a lifetime of wealth stashed on this ship!"

Tanya stood up and got in Tony's face. "Did you not hear what Jemma said? Your Dad is one of those about to attack a Battleship. If that Cruiser covers the Battleship, your dad and mine are dead. Is that worth a lifetime of luxury? I cannot use the word comfort because I would be so guilty that I did not save my dad. I could never rest."

Tony recognized his poor choice of words. "Ok, Tanya, you make a lot of sense, and I know how to get close without being molested. Assuming your Dad does not go out of his way to put another notch in his ship."

"What are you talking about, Tony?" John asked

"Simple, we hollow out a small iron-nickel asteroid and make our rogue asteroid that runs a course near the Cruiser but is not close enough for the Cruiser's crew to initiate their defenses. While on the way, we release the rest of our explosive chemical bots that we have modified with little engines. They are too small to be a concern. We give them the cruiser's missile tubes and engines as targets. After these, our little friend's attack, we play old Earth Trojan Horse, blow a cap we rigged on the asteroid, and come out fighting mad!"

"Yes. That would calculate a high probability of success if we disable the reactor cores or destroy the engines". Jemma interjected.

John added. "Can we then roll the asteroid into a collision course with the battleship? I want to give them another target other than a mining sled."

Sue then added. "I believe these people murdered my parents. I don't know how, but I believe they did it. I am in! Even if I have to drive a rocket into that cruiser myself."

"Ok, please update Isaac and the miners that there will be an extra visitor - a rogue asteroid with a mission, please, Jemma," John asked. "Oh, you might do that with your character Golden. I do not think talking to the parents is a wise move. Anyone object?"

Tony was now putting on the immersion headset. "We need to find an asteroid near the exit point where we believe the Cruiser will be patrolling. We must be far enough away to hollow a proper compartment without being seen. Our course needs to be a naturally smooth, predictable course. They must see us early to warn of a rogue asteroid with a predictable course. They will have other things to worry about and flag us as something not to fly into."

"Sue, you and Tanya are on the rotation for pilot and copilot. Tony and I will start looking for the right asteroid. I assume we are taking the Jungle route, which will get us within easy reach of the station. It would not be a surprise if the rouge asteroid emerged from that area.

Tony now had the halo activated to the 3-D projection near the station. Some minor asteroids started to light in the tank. Jemma had plotted the most probable primary courses for the Mantle Works fleet. She also posted two alternative approaches, P30 and P60 outcome ranges. Tony was already focusing on the area where Jemma had posted the course for the cruiser. The ship's role would be to render aid to either the destroyers or battleship if they needed it.

Tony excitedly said, "This will be like chopping down a Christmas tree, which I've always wanted to try. We'll burst out of the asteroid like a tiny superhero mouse popping out of cheese to face an evil cat."

Chapter 26
Cirrus Minor - Mr. Golden – Trex – Orbit below Station

The world has known many reclusive geniuses. Nicolov Tessela was one of these incredible scientists who changed the world – quietly. Generally, intellectuals passionate about weaponry end up in the Military Complex, often guarded and held as national assets. The national asset is placated with fun toys and a budget for their work. They are also effectively incarcerated with limited access to the world. Los Alamos was initially set up as such a research facility. The intellects came from Europe at the end of WWII. They had a choice. Live in this lovely town, work, live your dream, or face years of trials and torture due to your science. Indeed, the spoils of a global conflict went to the winner—a genius who saw man's track record of treating highly intellectual people as abysmal.

Trex had seamlessly slipped into its preplanned orbit around the station, rendering it challenging to detect without specialized probes. The station, ever vigilant, had deployed mining sleds as an improvised defensive canopy. These sleds were scheduled to rejoin the station upon the anticipated arrival of the Mantle Works fleet. As miners responded to the recall, the station buzzed with activity, focusing on upgrading their pulse cannons to a newly devised configuration. Additionally, a number of sleds were being outfitted with external missile racks, each bearing six anti-ship missiles. Isaac, having supplied seven hundred spare missiles, found himself wishing for an additional thousand, especially given the looming threat of the Cruiser. He weighed the option of diverting half of the missile-equipped mining sleds to confront the Cruiser rather than targeting the battleship. Yet, the battleship's formidable defense layers—comprising fighter support, anti-missile systems, close-range lasers, and depleted uranium Gatling guns—required a considerable missile barrage to breach, a feat yet to be accomplished. Isaac pondered a dual assault strategy to overwhelm its defenses but remained resolved to address the Cruiser's threat without squandering his valuable missile reserves.

"Commander, A call from the guild master and his special envoy?"

"Good timing. I was thinking about him. I will take it to the ready room." Isaac walked through to their ready room.

"Good afternoon, guild master. I hope the preparations are going well?"

"Yes, and I think the solution to our stray Mantle Works Cruiser has just come from one of our guild members currently off-station."

"Interesting, that has indeed been bothering me." Isaac was frantically waving to Mackena. He wanted her to hear the conversation.

"I would like to bring him into our conversation."

Isaac switched the call to the halo screen as Mackena walked in. The halo view changed to a small picture of the guild master with a larger image of an individual in a hooded robe and that same deep male-tenured voice. "Good afternoon, Isaac. Do I refer to you as commander? Also, madam Mackena, your respecters are grateful to you. The hooded man bowed deeply."

Isaac could see the inside of what he thought was a mining sled, but this one was the make Isaac had never seen. They saw two unidentifiable people in helmets rapidly adjusting a central halo projector. Asteroids were highlighted and then color-coded.

Then, the deep, gravelly voice slowly began, "I am known as Golden, but anything you want to call me will work. I've recently been called worse." The figure under the hood took a deep, deliberate, raspy breath as he raised a metallic arm with six digits and pointed two at Isaac. It continued, "I am aware of your Trex and its capabilities. This mining sled has been seriously modified so I could fulfill a promise to ensure the safety of its occupants." The hand dropped as the hood shifted slightly, revealing a bright golden reflection around the outside neck of the Golden cloak. The letters CTR could be seen inscribed on a gold necklace. The creaky old character continued. "We have modified this sled and trained for an offensive attack if needed. They have enough training and firepower to take that Mantle Works Cruiser out with room to spare. We understand all your capabilities."

"How do you know what my ship is capable of?" Isaac replied

"I scanned it when you visited the station, and yes, you did have it shielded."

"Good, who are you, Merlin, the Magician?"

"No, darker, much darker, "the man chuckled and paused, "just an old dog with a few new tricks. Here is what I thought might interest you. We can take the cruiser out of the fight without you using any of your valuable resources. You will need them against the battleship."

"Well, that sounds a little bold. Are you perhaps named David?"

"No, not a bad suggestion for a middle name, however." the man replied again with that subtle laugh that Isaac could not figure out if it were evil or just a person with a good heart who liked to make light of serious situations. Still, that deep guttural laugh bothered him.

"Go on."

"We have hallowed out a small iron-nickel asteroid large enough to hold the modified sled. This will enter the battlefield as a rogue asteroid on a trajectory that will take it close to the Cruiser, but not a clear threat."

"Interesting avoiding a missile dual, perhaps?" Isaac questioned.

"You're right. The sled has weapons and can defend against missiles, but a cruiser could be too much for it in a direct fight. It's better not to risk it. As we head towards the cruiser, we'll keep an eye on its position and speed using the passive sensors we've set up on the asteroid. We've also modified some mining equipment called mine bots. They now have an explosive charge and a very strong acid that eats through metal. They can move a little on their own but don't go very far. As the asteroid spins, we'll send these small mining probes towards the cruiser. If the cruiser hasn't turned on its energy shields, the probes will land and look for weak spots like missile tubes, airlocks, and service ports. We've got a little over a thousand of these gadgets ready.

If the cruiser's shields are up, the probes can sneak in close to the engine exhaust and the shield to hit the engineering areas. We think the acid could do serious damage to the engineering bay, turning it into a dangerous mess."

The Guild Master observed, "This is an interesting twist. What about the sled?"

"We will release the sled just past the perigee with the Cruiser. If their captain is suspicious, they expect an attack before the rock reaches its closest point. Remember, we will be in the cruiser's primary plasma and Ion canons range but not inside the secondary weapons envelope. We don't intend to ever be in the scope of the ion cannon, which is forward-facing.

The moment we release from the Trojan horse, we will launch a set of probes that will emulate a set of destroyers coming through the asteroid field and the pink puffballs. The pink puffballs are the same defense modified from you, Isaac, but this is on steroids. We were able to conduct some critical testing recently. We hope to see the Cruiser drive into it. It will reduce the impact of the Cruisers lasers and pulse cannons if necessary. The alexandrite lasers have also been tested recently. After a recent encounter with two ship-killing rogue asteroids, we adjusted these for higher output. We will use these to target the cruiser's engines. Three of the plasma cannons will target offensive weapons. We have a mass accelerator and five shots targeting the main bridge and deeper command center. Since these are earth-designed ships, we have a perfect idea of where to place these bullets. We will do a high-velocity pass, giving us enough time to launch five rounds from the mass accelerator. This sled will surprise you with its speed and maneuverability. Commander, let me assure you this is not planned to be a suicide mission. This sled carries a very special load, and we will avoid destruction at all costs."

"Seems pretty risky," Isaac commented.

A young officer was projecting a model from his wrist computer and got excited, "Sir, Sir." He was attempting to get the senior elder's attention. The elder turned and acknowledged the young man. "Sir. With the Cruiser out of the picture, you have a ninety-seven percent chance of success. Without it, fifty-one percent. Sir, Mr. Golden is right!" A young miner was reading his computer model.

"Anything we can do to help you. Something like a few missiles heading for the Cruiser?"

"No, the Cruiser must think it is clear and has no threats. Landing the Chembots increases success. Our rating passed ninety-eight percent."

"Anything we can get you?"

"Yes, please announce to your sled drivers that you have hidden in the near field asteroids that a medium-sized rogue asteroid is not a target. Some of them will go out of their way to attack rogue asteroids. We do not need that distraction, which would alert the Cruiser that all is not as it seems." The dark gray, gruff older man asked.

Isaac was shaking his head, thinking this was an extraordinary genius and very eccentric. He had to meet this one incredible creature, dark or not.

The guild master then said, "I will handle that. I know who you are talking about."

Isaac asked, "I have now had two meetings with you, and I do not know who you are."

"We have already met, Commander. Maybe we will meet again after this issue is handled. See you on the other side."

Chapter 27
Cirrus Minor - Battleship MF

Building and maintaining war-based vessels for space was nearly prohibitively expensive. The fleets were small and used only to enforce a legal claim or delay activity until settled. The Battleship MF was the only one of its kind. It was used around the Mantle Works Asteroid mining complexes in the asteroid field in Earth's solar system of planets. It was busy and had only been used as a heavy-hitting threat to remove obstacles. It had never fired a shot in anger.

The battleship, cruiser, and three destroyers were now six hours from the station. Allison had called a strategy meeting. She wanted to clarify that the battleship and Cruiser were to be high-level enforcers. The destroyer's job was to take the station with troops.

"Captains, I want this evident. The primary mission is to capture the mining station intact. We cleared the defenses last time we were here. They were insignificant. The guild may have shored up some minor defenses. They are fundamentally a bunch of slackers, and we are here to help. This mining facility is valuable to our future. Once captured, we will teach our laborers what real productivity is. Trooper commanders, this means squash those rebellious with extreme violence. All doctors, teachers, religious leaders, and researchers are considered rebels. They are educated. We no longer want this population to think. Do you understand?"

273

Five ship captains and three troop commanders acknowledged Allison's words almost verbatim.

"Good, you will have fun at Asteroid 16' I believe these rebels call it. We will use the battleship and cruiser to show the miners what extinction looks like!"

Allison turned to Chadoom, who was clad in a replica of the Seventh Cavalry general's uniform from the Battle of the Little Bighorn in 1876. His outfit was striking, featuring a bright gold shirt and a royal blue jacket accented with gold piping, complete with a double row of gold buttons down the front. Around his neck, he wore a red silk scarf. However, he was missing the wide-brimmed, flattish hat that would have completed the ensemble.

Allison tossed a grenade at Chadoom, "Really, Chadoom, is this the right message with the glitch uniform."

Chadoom smiled and said, "Somebody has to lead this group. I like to do it in style."

"Are your fighters ready for missile defense? That was the one thing we did not knock out last time we were here. I want them clearing the path for the destroyers."

"Yes, mam, we are ready. That base will be ours. The battleship will take medium orbit with the Cruiser on top looking for bandits from above."

"What are you talking about, bandits from above?"

"A sled attack, but we expect the miners will be disorganized, and attacks will be sporadic and inefficient as these miners are."

I am assuming their main armament is a plasma cannon, correct?"

"Yes, but it would take a couple of hundred mining sleds to get near the cruiser or battleship to do any damage."

Allison looked at the Cruiser captain. "Are you ready for the top cover?"

"Aye, we are ready, mam," replied Captain Turner.

"We will cover the destroyer attack and troop landing. Remember, our goal is to capture the station, not destroy it! Now, who has an intelligence update on the station?"

"I do, mam." said a young lieutenant," We have a minimal view. Our normal sources of information have dried up. Commander related to armament, Chadoom had a little intelligence that the miners had been working on their defenses, but nothing we could substantiate through our normal channels. We have sent three probes ahead of the fleet. While each probe was destroyed, we can confirm the anti-missile defensive lasers have been replaced. We also believe they have jury-rigged a force field, as our optical sensors could not get a top view of the station. However, dark patches are seen and consistent with where we targeted the plasma cannon and laser emplacements several weeks ago. We concluded the mining guild has not replaced much of the armament we destroyed."

The battleship Captain Henry broke in and said. "Typical mining laziness. They just don't care. They are in this for their profit, and the guild is a bunch of useless older adults who promote their wicked ways. This is a walk in the park. It is a nice break from taunting Earth forces. At least Earth is a nearly worthy opponent."

The young lieutenant spoke a word of caution. "Mam, we don't know how much the guild has rebuilt its defenses. I would have focused on missiles and missile defenses if it were me. We can guess they have done the latter because missiles are easy and cheap to build."

"Very well, lieutenant, off my bridge, please," the gruff old captain ordered.

Allison looked at the captain. "Remember Henry. You were one of those at one time."

Chadoom stepped forward. "The destroyers will trail the new Type One fighters. These fighters will eliminate missiles and clear any obstacles in our path. They will maintain their course until our troops are securely landed. Their secondary objective is to neutralize any vessel attempting to escape the station. Destroyer captains, I want you to capture video footage of the fleeing sleds being destroyed. It's crucial that the new subjects understand they are under the control of a new ruler."

"Good. Thank you, everyone. We will complete the job Mantle Works should have finished in the Earth's asteroid belt years ago. Dismissed!" The room emptied, leaving Captain Henry, Allison, and Chadoom.

"Chadoom, is our stray controller still on board," Allison asked.

Allison's right-hand man winked at the captain while wickedly grinning and reported, "I am sorry, Allison. He got confused over how to operate an airlock. I regret to say he accidentally walked into the airlock and opened it before I could save him."

The Captain stepped in. "My maintenance crew reversed the controls during routine maintenance the day before. I have dealt with the two techs. Terrible loss. He was about to audit the food storage."

"Well, that solves another mystery. Let's go to the command center and monitor this operation."

Chapter 28
Cirrus Minor - Rogue Asteroid- Mining Sled 2

The account of the Trojan horse in 1184 B. C. brought to light how clever the human mind can be in setting up a trap. This bit of lore demonstrates the actual hunting genetics of the human race. It is another deadly strategy in the violent game of war.

Mining Sled Two was neatly packed inside a hollowed-out iron-nickel asteroid. They had hollowed out an engine port on the opposite side of the asteroid to provide early thrust to the asteroid, which was sitting three hundred and fifty thousand kilometers away from the open space surrounding the station. They did not know precisely where the cruiser was but selected the highest probable area where it would patrol based on their modeling. Sue and Tanya did most of the piloting while John, Tony, and Jemma modified the chembots. Jemma had reconfigured her arms, so she had dexterous hands instead of paws. She thought paws just made life easier. She was blazing fast with her nimble six-digit hands.

"Jemma, your man imitation hologram of Mr. Golden is good. Have you been watching old Earth science fiction movies again? It's something like the show called Star Wars, where they have the evil emperor who always wears a hood and talks in a low, groveling voice. That was priceless—David as a middle name. I thought you were getting us the reputation as giant killers, too. Let's see a mining sled taking on a Battle Cruiser. That is just suicide. But here we are, one happy, wealthy family ready to take one for the team."

"Technically, Isaacs's ship is a battlecruiser. The mantle Works is a cruiser."

"Did you make the cap to cover us for after we accelerate this big iron ball? Once we are in detection range, we'll need that cap," John said, putting on his hard shell. His job was to put the plug into the thrust hole once they had accelerated the asteroid.

"Yes, and the plug is on a gravity-sled. That will make the work fast. You will be subjected to some nasty conditions if we get any rotation. Like puke in the helmet stuff." Tony enthusiastically added.

"Thanks. Where are our friends anyway?" John asked.

"The station master is beaming a continuous signal informing all parties of Mantle Works' positioning. Hard to hide when you drive massive trucks around." Tanya informed the group. The cruiser is fifteen minutes out from entering orbit. Five minutes before we launch the probes."

Jemma chimed in, "The guild deployed a few mining sleds, positioning them to fly just out of weapons range, both ahead and behind their targets—clearly a tactic to make them feel pursued. The battleship then launched twenty fighters, whose flight patterns and formations suggested they were AI-controlled. I suspect these fighters will spearhead the destroyers' assault. Once we've dealt with the cruiser, we might have to hunt down any remaining fighters. They pose the greatest threat to our future operations."

Tanya spoke up from the CO-pilot's seat. "What does that make you?"

"Very dangerous to them. Remember, I have weapons of my design."

"How dangerous are you to our future?"

"I hope I am proving I have a conscience and want to be part of your future and your children's."

"Good answer, a real ministering angel dog."

"I like that," Tony piped up, "an angelic dog with an attitude."

John announced, "Ok, time to focus!"

As Jemma updated the situation map, the halo display transitioned from a projection to a live view. She briefed John, "The cruiser has assumed a high orbit, just as we predicted. That's our target. We've plotted a course that puts us in an optimal trajectory—close enough to launch missiles but beyond the reach of their plasma cannons. We're not even worth an ion cannon shot. John, it's time for you to get acquainted with the airlock. Everyone is suited up in at least soft shells, with hard shells on standby. The weapons station is secure within its protective bubble. Tony, you're up first."

Tony settled into the newly designed cocoon with a cover resembling plexiglass, which he smoothly slid closed. Jemma and Tanya had engineered these emergency survival shelters around the immersion stations that housed the sensors and weapons. Tony updated the team, "We've got a passive sensor lock on the cruiser. It's emitting a lot of signals. We'll shut down all non-essential systems, keeping only a minimal situational feed active once we're underway. Life support will be minimized, too. We're running silent." Tanya then took over, announcing, "Starting thrust checklist; prepare for engagement. We're in this to win."

"Spoken like warrior princes." Tony jabbed back.

"We start releasing our little friends from the surface three minutes after we complete the burn, right?" Sue nervously reported.

"Airlock depressurized, ready to go. Bots are hot." John said.

"Jemma, good to go?"

"Yes, if those fighters leading the destroyers come this way, we will eject and fight through them first. They have no life forms and, while not powerful, are deadly. I will fly the ship if we have more than two incarnate pieces of refuge on our tail." Jemma announced

"10-4 something does worry David after all. That means it's serious."

"Only if any of them survive," John said, exiting the airlock. "Tanya, give me a second to get secured." He locked a tether onto a peg on the side of the ship. "Go for it when you are ready."

"Drive on three, two, one," Tanya announced

The ship shook as the asteroid started to respond. It was a shotgun start with severe g forces. The rock was sailing true to the course. No electronic leaks were detected by their passive sensor array outside the asteroid.

"Ship is on course and flying true. We are at speed. John, you have three minutes before we are detected. Destroyers are in a position to start their run on the station in about four minutes. Jemma remembers Mantle Works fires the first shot. Let us know when that occurs." Tanya summarized.

"They just did. Fighters destroyed a mining sled with a family vacating the station. They had surrendered and were powering down when the leading fighter gunned it down."

John detached his tether and navigated the narrow space between the sled and the asteroid. A nickel-iron plug lay on the gravity sled, along with a tube of instant sealant for emergency repairs. He took hold of the gravity sled controls, carefully positioning the plug into a hole with glowing red edges—superheated by the ship's exhaust. Although the plug fit, it required some additional filler. John promptly filled the gaps with the pliable, heated rock, using it to cement the plug in place. After securing the seal, he began returning to the sled. A voice crackled over his communicator, "John, a small problem. We've lost communication with the pods. Can you check it out?"

"I am on it now. When do we need to launch?"

"Now would be nice," Tony retorted

John followed the cable to near the airlock, where he saw a ceiling fragment had come loose and severed the cable. He immediately pulled his wire cutters out and shaped the cable leads to plug into his wrist computer like when they rigged up mining charges. It took about thirty seconds. He was sweating when he plugged the manicured end into his laptop and sent a single-word command, "Execute."

The lead probe sent back acknowledged, and he detached the cord. Tiny probes started to rise from the asteroid's surface a few at a time in a random sequence. The small jets the probes used were not detectable even to the passive sensors outside on the asteroid's surface. John could not use the radio due to the radio silence. He also manually manipulated the airlock. Once entered, he went straight to the weapons station. He high-fived Tony the best they could through the protective shields when he settled down in his cocoon.

Chapter 29
Cirrus Minor - Cruiser's Folly - Mantle Works Cruiser

Preparation for the battle is the most crucial element to win the battle. Over-estimating the enemy in any capacity is a recipe for disaster. Mantle Works had built a navy, but years had taken a toll. Now, they were filled with prejudice, which caused them to make terrible decisions.

"Captain Turner, a rogue asteroid has come out of a dense asteroid cloud. It is no threat to the ship. It will pass clear of all ships in the fleet. No threat."

Turner responded. "Run a full scan of the asteroid, please, Lieutenant."

"We have no detectable electric activity and slow lateral rotation, sir Nickel-Iron. Plenty of collision evidence." The lieutenant leaned his ear to one side. "Intel also confirms Rogue Asteroid - no threat."

"Probably a miner was trying to push a rock at us, hoping to get us off guard. Any more of these rogue asteroids heading our way."

"Negative, and we checked with the Battleship MF sensor group. They report all is quiet for now. "

Captain Turner spat, "Stupid Miners! It is not even worth a missile."

"Sir, Message from Commander Chadoom"

"Yes, what is it."

"Attack commencing"

"Keep the ship at ready alert until we have a culpable threat. If we have to save anyone, we go to battle stations."

"Aye, Captain Turner," the lieutenant answered, relieved that they would not be on Battlestations for a while if this went well. Battlestations were always tense times. You would check your weapons or review your electronic sweeps, just waiting for a few seconds of absolute terror.

Chapter 30
Cirrus Minor - Trojan Horse - Cirrus Minor
Mining Station

Wars in space are mysterious ballets of fast spaceships on a gigantic playing field. Spacebattles, no matter how exciting they might seem from a distance, are nothing more than a few minutes of terror. Indeed, tragic events result from intelligent (?) beings directing untold energies at one another. Was it any different from the knights in shining armor whacking at one another with big chunks of sharpened metal?

"The Guild Master issued a decisive command: "Fire the station missiles. Execute Plan Alpha. Ensure the missiles are launched beyond the range of electromagnetic weapons. Hold the Alexandrite Lasers until the missiles are clear. We want their forces to be so deeply committed that there's no turning back."

"Lasers, missile defense, pulse cannon, and Ion cannons are ready."

"Fire the pink panthers for incoming fighters. Let's see how they react."

One of the miners manning the radar arrays announced, "Destroyers have launched missiles. Count sixty missiles inbound. The battleship just launched ninety more inbound missiles."

The battle screen in the mining centers showed the sixty inbound missiles from the destroyers and the ninety following from the battleship, separated by fifteen degrees and would impact twenty seconds later.

"The fighters moved behind the first wave of missiles. Missile impact in five minutes. EM-Bomb in four minutes and thirty seconds. The yield on enemy missiles is low, one megaton. They want to shake us up and make sure or take our defenses down before they board."

"Sir, fighters have destroyed eighty-five percent of our missiles—four are clear. Presently, the inbound target is destroyer number one, "The miner turned radar technician announced.

"Is the shield ready to activate?"

"Yes, the shield is ready, on standby."

"Time to first laser shot?"

"Ten seconds, we will focus on the first destroyer. Destroyers are three minutes out. Battleship has launched fighters. But in a protective pattern."

Near the Mantle Works Cruiser:

The rogue asteroid had traveled through its perigee to the Cruiser while the tiny chemical bots had flown in a thin stream toward the cruiser, making minor course corrections. Over ninety-five percent landed on the cruiser's hull. When the rogue asteroid was at perigee with the cruiser, the bots were like mosquitoes thirsty for blood. They moved to their assigned locations around the cruiser, waiting for their timers to allow them to execute their final duty. One-quarter of them locked their shaped charge and blew a hole into the cruiser's skin, pouring their corrosive metal-eating acid into the ship. They waited for any atmospheric release from depressurization. Then, the rest of the chemical bots entered the ship following their predecessors.

All the internal alarms on the bridge in the Cruiser were triggered at once when the inner pressurized hull was penetrated. The central alarm announced, "Intruder alert, Intruder alert!" with two very loud beeps repeated repeatedly. Captain Turner had just settled after inspecting each of the bridge stations. He enjoyed his favorite pre-battle drink - cinnamon hot chocolate with whipped cream. He jumped when the alarms sounded, spilling the drink on his freshly pressed uniform blouse after hitting his mouth on the edge of the cup. He was now wearing the whipped cream as a mustache.

"Kill that alarm," he trumpeted, wiping the white mustache off his face. "Report! Battle stations!"

"Sensors offline, Decompression all decks, eighty percent of the missile tubes are offline." The ship shook with a significant explosion, setting off another set of alarms.

"Intruder defense teams now! What hit us?"

An urgent report came through: "Sir, missile bay fifteen has suffered an internal explosion. I recommend we suit up for decompression immediately. Fifty percent of the ship is now without air. Additionally, we're unable to locate any intruders. The only signs are corrosive chemicals and minor explosions throughout."

"Get those sensors online." The captain slapped a button on his council. "Engineering, what is our status?"

"Suited, No atmosphere, but we are still one hundred percent. We can fight. Private, what is that thing? Oh crap!" a loud bang was heard over the speaker. The lights dimmed.

"Engineering report!"

"Sir," the bridge officer said, "We just lost reactor number one."

The captain hit another button. "Commander, get my ship clean of these intruders or whoever they are - now."

"Bridge engineering," the chief engineer was bleeding and covered in white blemishes from acid still burning on his skin. He was in pain, and he sounded hurt. "Lost reactor one and number two is dropping like a rock. Three and four are at one hundred and ten percent. Your intruders are small automatic robots. They have what I think is a shaped charge and a chemical corrosive. Very nasty!"

The Captain ordered, "You get this, commander. Find this nasty little beast and get them off my ship." Then Turner bellowed, "Someone, I need an outside visual and use the nose radar to get me a picture."

The main screen flickered to life with a one-hundred-and-eighty-degree view.

"Patch me into the Battlespace sensors from the Battleship. We need to know what is going on!" Turner ordered.

"Aye, Captain, channel open," responded the communication officer.

"Captain Henry, we have an issue on the Cruiser. Captain Turner reported to his associate on the battleship. "Intruder attack, fifty percent of the reactors are down, eighty percent of the missile tubes are down, all external sensors out, patching into Battlespace for sensors. Visual and forward radar is up. We can fight with pulsed weapons and lasers." The ship's hull shook hard again, forcing the captain to grab a nearby railing for stability.

"Sir, Plasma cannons two and three are out. Ten percent of the anti-missile lasers are out."

Turner raised his voice in frustration, "Would someone shoot these things?"

Mining Sled Two:

Two minutes after the closest approach to the Cruiser at the height of the chemical bot attack, the cap on the asteroid silently blew out. Mining Sled Two (renamed Mighty Mouse after an early Earth cartoon by Tony) popped out from within the metal asteroid. Mighty Mouse, with the modified engines, accelerated hard toward the cruiser. The Trojan horse had revealed its deadly cargo.

"Firing pink powder puffs for missiles. Deployment in ten seconds."

"Firing Lasers and the first shot of the rail gun in ten seconds," Tony announced

Sue announced, "Evasive course A is being flown. Visible breach and damage to Cruiser."

Visible pink clouds formed before the Cruiser, creating a large cloud. The Cruiser would soon fly through it. The purple beams of the alexandrite lasers lit up the space as the two bolts impacted the Cruiser and sliced right through to the other side. The metallic bullets from the railguns followed the laser strikes. The two shots had small entry holes but massive exit holes with significant debris pouring out of the ship's side.

Mantle Works Cruiser:

Back on the Cruiser, things were settling down after the deck had buckled under Captain Turner's feet as the ship shook again.

"That was not a missile tube exploding! What is attacking us? What is that Pink Cloud? Get my shield up. That is your top priority."

"Sir, the communications are gone. And internal ship communication is at twenty percent."

"Damn, get me some shields," he said as the ship began to rattle, hitting the metal shards comprising the pink cloud.

Sled two (Mighty Mouse):

Tony said, "Second laser shot in three, two, one, now, rail-gun shots three and four. Target engineering." Again, the purple beams lashed out and hit the rear portion of the cruiser. Again, like the first strike, the beams tore through the cruiser, followed by two other impacts, which created massive exit wounds like the first. This time, secondary explosions ripped through the rear of the cruiser. The engines went black. The cruiser was dead.

The second shots tore the reactor cores out, saving the crew from reactor meltdowns and explosions.

Mantle Works Cruiser:

"Captain Turner, No response from engineering. We are on backup power only."

"I had that figured out when the lights went out. Go get me a visual report of the damage."

"Sir, battlespace showed a single sled approaching the cruiser before the strike."

"What, we have a single mining sled doing this amount of damage?" the Captain stared at the junior officer.

The ship started to rattle even more with additional explosions.

"Sir, Weapons are taking plasma cannon hits. The main plasma cannons are offline. The battery's energy is being drained back into emergency batteries. Engines two and three have been surgically removed."

"Lieutenant, signal our surrender. We are done. We are dead, abandoned ship".

Battleship MF:

Back on the battleship, Chadoom was monitoring the battle at the mining station. Fighters were in position, missiles ahead and behind them. They would strafe any remaining defenses just after the first missile strike. Then, the fighters would fall back and let the second set of missiles punch the station for the landing parties.

He then noticed the cruiser had fallen and dropped off the board. "Chadoom to Captain Henry, what has happened to the cruiser."

"Captain Turner indicated he had an intruder attack of little corrosive chemical robots. His communication officer then suggested they were under attack from a single ship. We have spotted only a mining sled, but that is beyond the capability of a sled."

"Captain, send ten of your fighters from local missile protection over there and clean out anything still bothering Captain Turner," Chadoom ordered Captain Henry. Captain Henry did not like being ordered on his ship but relented and ordered thirty seconds later.

Cirrus Minor station with a hardened link to the TREX:

"Electromagnetic Bomb in thirty seconds, system shutdown in fifteen seconds. Are the gunships under the EM protection barrier?"

"Aye,"

"Commander Isaac, we are ready for EM. The Mantle Works cruiser is down, as promised. Ten mantle works fighters on the way to aid the cruiser, and that little fighter has the profile of a mining sled and the power output of a cruiser. Miners are lining up a Mining sled missile attack on the battleship."

"Don't worry about the mining sled by the Cruiser. I think they can take care of themselves. Good luck. If we trap those twenty leading fighters with the EM, go after them first before they can recover, then we hit the next wave of missiles from the Battleship."

"10-4, seven, six, good luck," and the link was dropped.

The Guild Master started to pace the dark floor. The main hologram screen was blank, and the base would be quiet for fifteen seconds. If this did not work, they would have enough time to fight the first sixty missiles, but the next ninety from the battleship would be challenging. In addition, they would still need to fight the destroyers. "Guild Master, we are ready to repel invaders," a guild member reported to the station's leader.

"We have two hundred and fifty miners in hardcover ready to fight. They are armed with the new handheld plasma cutters. Remember, they are protecting their homes."

"EM bomb in three, two, one." Everything went black on the station when the count hit one.

Eleven seconds passed by before the halo screen came alive.

"Sensors, update the situation, please."

"Status on incoming?"

"The first wave of missiles is inert; the fighters are dead and coasting because they have no power.   The second wave of missiles behind the lot is at normal burn for another five minutes. The lead destroyer is drifting but may be ready for a shuttle launch. Destroyer Two is at a reduced speed eleven minutes out. Destroyer three 13 minutes out and accelerating. Battleship same course and orientation, ten fighters are approaching the cruiser and Golden's sled."

"Launch gunships to clean out those incapacitated fighters. Plasma cannons hit the same fighters first because they are the biggest threat. Lasers target the third destroyer. Launch our pink powder puffs. Put them just behind the fighters. Missiles away at the third destroyer."

The holographic screen continued to update as thirty gunships launched from the station and another twenty missiles targeted the third destroyer. The first drifting destroyer exploded when the two missiles launched from a mining sled tore into the front of the drifting destroyer. These dissected the destroyer in half. The two assault shuttles in the process of launching at the time of the missile intersection drifted back toward the ship. They crashed, bounced, and spun away in flames, filling space with escape pods from the troop-laden assault shuttles. Two others continued to drift on a course for the station.

The gunships joined the battle and started to execute the Mantle Works' experimental AI fighters. The emp blast had been powerful and very effective on weakly hardened electronics. The Gunship line efficiently dispatched the seemingly dead advanced Mantle Works fighters when the last two recycled through their restart programs.

The quick, agile fighters tore open space to loop around and attack the gunship that was now dead ahead of it. The gunship disappeared as the other fighter pulled into the line of gunships. The mining station's gunships had wiped out fourteen AI soldiers and severely damaged two more as the gunships continued their fight. A lucky laser shot destroyed the fiftieth Mantle Works AI. Fighter, as the two AI survivors ripped through the gunships, killing twenty in the first pass. The gunships were too slow and poorly armored to stand against the intelligent, well-coordinated fighters. The two damaged fighters then turned and started toward the battleship. The ten remaining gunboats sought safer shelter and attacked missiles inbound for the station. The uninjured AI fighters turned and joined their injured companions as an escort.

Meanwhile, the station lasers impacted the second destroyer, cutting the ship's main plasma cannon in half. The destroyer immediately began to twist and turn to avoid any more long-range laser strikes. When it reached the outer limits of its range, the destroyer brought its lasers online, which bounced harmlessly on the station's shields.

As this happened, the mining station allied fighter cruiser moved away from behind the station. The fighter detached as the Cruiser lined up for a missile launch at the battleship. Simultaneously, two large groups of mining sleds moved out from two large asteroids. One hundred of the sleds had missile pods attached to them. The battleship's remaining fighters formed on the battleship to provide an outer shield.

Battleship MF:

"Captain Henry, prepare for an attack run on the Mining Station," Chadoom commanded.

Allison was worried too many things were going wrong. She had lost a Cruiser and a destroyer, plus sixteen of Chadoom's artificial weapons. "Can you tell me what these weapons are and how the miners have weapons we have never seen? So they have shown us what they have now; let's get this job done."

The ship's Captain replied, "Yes, mam. We are on it. I have sent Lieutenant Velma down from Intel. She has some impressions."

"Well, get her here now, Captain!"

"Chadoom, what is happening with your A.I. Fighters? Four are returning - why do we still have station missiles inbound on the destroyer, two reporting damage and losing fifty percent of their shields."

"Allison, this is a preservation program.", Chadoom retorted. "The two functional fighters are protecting and escorting the damaged. The other two will join our defensive force when the damaged fighters land. This is programmed after a fifty percent force loss."

"That is one bit of programming I will have you change. Once on a mission, they must fulfill it at all costs in the future!"

Lieutenant Velma rolled into the command center. She was in a hover chair with an oxygen line supporting her breathing. She had white hair and looked ancient. She was focused, however. "Commander Chadoom and CEO Allison, I was asked to inform you what weapons we see."

"Proceed. You look ancient enough to be my grandmother." Allison chided.

"The opposition is using traditional missiles and modified plasma weapons. The pink clouds we see have metal shards that will tear our missiles up. Their laser is unknown to us and has at least doubled the range and quadrupled the power. We have just analyzed the weapon they used to knock out the first wave of missiles and the fighters. It is an electromagnetic weapon with a conical shape impact area. The station shut down ahead of detonation. If they use it again, we will have limited warning. That mining shuttle that knocked the cruiser out is an unknown ship with the shape and size of a mining sled or an Earth-based warship. There are ten fighters on the way to sort them out. This ship ambushed the Cruiser. "She finished and breathed hard, causing the oxygen pump to rattle as it worked harder.

Allison complimented her, "Thank you, Lieutenant Velma. That was concise. Where did the station find the new technology?"

"We believe they invented it. We believe they had to have developed these weapons just after your last visit," the older Lieutenant indicated.

"We did burn their defensive weapons down," Allison said, looking at Chadoom.

"Thank you, Lieutenant. Oh," he said respectfully, "Go in peace."

Bridge of Isaacs fighter (T):

"Mackena, we will launch ten seconds after the Mining Sleds," Isaac reminded his best friend. "Please remind the sleds to move to rendezvous point 'A' post-launch."

Mackena replied, "Sleds are now firing the first 100 missiles. Our missiles will launch in ten seconds, and we will fire our wing mounts plus mains. After that, we will follow up with another hundred ten seconds later. Our missiles are five megatons, and the mining sleds are three megatons. It will take all we have to overwhelm that battleship's shields, assuming a sixty percent loss of missiles. Isaac, if this fails, you must take the shield out with your main gun."

"10-4 ready," Isaac said, adjusting himself for the umpteenth time. This was no simulator.

Mighty Mouse (Mining Sled Two):

Jemma was ready for action but was still concerned about breaking her CTR code. She felt the three robotic laws would be helpful in forcing her to do as little harm as possible while saving her people, but she would not be choosing her course. No! Being a free-thinking individual came with a cost of a conscience that no laws could duplicate. Endless memories, good and bad memories. Her job was to save her people – all of them. "John, Tony, Tanya, and Sue, I would take flight control and lasers with your permission. We have ten inbound fighters. In addition, there are four additional artificial intelligence fighters on course to the battleship. Two are severely damaged."

John replied," Tony can handle the main plasma cannon. I will control the topside plasma pulse cannon, Sue. Are you game for the bottom plasma cannon? Tanya, you handle railguns, auxiliary defensive weapons, and sensors?"

"Fine with me," Tony replied happily in his immersion capsule.

"We are returning to the immersion couches now," replied Sue. Sue and Tanya had driven the ship during the Cruiser attack.

"Jemma, ship control is now yours," Tanya acknowledged as she unbuckled her harness to return to the immersion couch.

Jemma was standing in her immersion area, driving the ship. Her tail was now between her legs.

Tanya stopped Sue as they entered the main ship area containing the immersion couches. "Sue, I am scared. Letting a computer control a ship and weapons has serious ramifications, which means we have a serious situation. This is what that silent prayer in the heart is all about."

Sue stopped with tears in her eyes, grabbing onto Tanyas' shoulders, "I am here, dear sister. I have been hanging on to you since my mom and dad were murdered. Now I am ready to forgive the evil one that executed them and stand like the warriors of old Earth to protect my family." She hugged her new sister, and they silently parted, made a fist bump, and moved to their couches.

They slipped into their immersion couches, which conformed to them, and created a safe cocoon as they immersed. Once in the immersion session with John and Tony, Jemma opened the little mining sled up and demonstrated its total capacity. The ten fighters entered the first pink cloud masses, thinking they would intersect the mining sled on the other side. Jemma increased the engines to eighty percent. The little sled shot away from the fighters' exit point. It swung in a tight one-hundred-and-eight-degree loop, coming about to target the fighters coming through the edge of the pink paint mixed with metal shards. Jemma fed continuous updates through the immersion network, which was complemented by Tanya's consistent sensor updates. The sled had much better sensors than the fighters, who were eighty-five percent blind to the little sled while in the cloud.

"Prepare to engage plasma cannon on fighters five and six," Jemma announced.

The fighters broke through, scarred from the shrapnel but still functional. The view screens were chewed up, making the fighter pilots utterly reliant on their internal radar and battle space connection with the battleship's Battle-space network. Unfortunately, this network was hindered by the cruiser being a scrap heap. The mining sled timed the intersection perfectly and fired four short bursts of the alexandrite laser at the emerging fighters. The forward lasers tore the fighters apart. Then, the plasma cannons hit fighters five and six. The main plasma cannon melted the engine off of Fighter Five. The combined fire of the top and base pulse cannons shredded fighter number six. This ship blew its reactor, catching Fighter Seven in the blast. It did not survive.

Jemma reinforced the forward screen and flew straight through the junk. The sled went between the cloud and the remaining three fighters, who were dazed because they could not spot the attacking ship. Jemma then conducted a similar move that put the sled into its first attack run. The fighters had finally figured out where to look and were increasing their speed to make an attack run. It was too late; however, the enhanced engines of the sled were too fast for the fighters to react fully.

"Plasma cannons share shots on nine and ten," Jemma calmly requested. One of the front lasers fired, cutting fighter eight in half. Again, the main plasma cannon fired. The cannon melted the front portion of Fighter Nine off, which caused the ship to start a fast spin toward the asteroids. The top pulse plasma cannon missed Fighter Ten because it reacted to Fighter Seven's destruction, but the lower plasma cannon targeted the fighter's cockpit. The pulsed gun cut a semicircular hole from where the driver would and cored the ship, leaving another debris field. As with the first group of dispatched fighters, the sled blew through the debris, rolling its wings in victory.

Tony cheered, "Hurray, ten out of ten, and do it in your home. That was so much better than a video game. What a rush!"

John was less enthusiastic."I missed my last shot."

Tanya and Sue emerged from the immersion couch, turned, and hugged one another.

Tanya said, "Sue, you are incredible. I was waiting for you to take that last shot. Wow, revenge, I would say. You cored the pilot out of that sled. What a shot! I was only thinking about where I would take them out. It is like you have extrasensory perception. You targeted the same spot. We must be like twin sisters."

"Ya, sisters with cannons must be a new hologram series. Something like an evil twin episode.."

John laughed. "Good shot, Sue. Does anyone have a direction for the pilot we sent into orbit?"

Tanya raised her hand. "Covered."

Battleship M F Command Center:

"Captain, get those two injured fighters on deck now. I will order the other two for missile defense. Three hundred missiles, where did they get those? Sneaky miners."

Captain Henry retorted. "Someone was blind here. I just lost ten fighters to that mining sled. I have lost one destroyer, and number two is compromised. Three might make it if we move the battleship now, but we have three hundred missiles bearing in on us. We will handle these, but it will strip our fighter cover, and without a doubt, we will take a little damage. I am having a terrible day. And Chadoom, your two injured fighters did not land. They are headed for the Asteroids."

Chadoom opened up his wrist computer and punched a button. "Captain, I have just taken care of the two fighters."

"Commander Chadoom, your two damaged fighters are still moving away."

"Hmm, that is interesting. I see the other two are in position for missile intercept." Chadoom observed.

"Commander, I am executing a down fifty-degree turn to the starboard for intercept on the station. We will have issues with missiles coming from that missile cruiser beside the station."

"Agreed, Captain. We will need to support the troops landing."

Allison butted in. "Correct, Captain, but do not use the mains on the station other than on the peripheral edges."

"CEO Allison, we still have another fifty missiles to unload on the station. I would like those to lead us in."

"That is fine, Captain. Just get me that station." She stomped out the bridge with Chadoom hot behind her. Allison stopped and put her hand on a door that scanned her hand and opened.

Captain Henry could be heard on Allison's wrist computer, "10-4 CEO Fighters are engaging the upper missiles, lasers are firing on the bottom missile cluster. This is going to get busy."

The connection to the bridge was terminated. Allison turned as she entered the tight cargo bay of a small ship attached to the Battleship. "My dear friend Chadoom, I have a bad feeling about this. This compartment is also an escape vehicle. Is that correct?"

"Yes, hit that big yellow button, and the two halves of the escape ship are sealed together. Each half holds an engine for quick escapes. We can be away in 30 seconds."

"OK, I hope we don't need it, but this day is not in our favor. And I have always learned to have an escape route."

The fighters were efficient and reduced the upper missiles by eighty percent. The AI fighters proved their worth and efficiently knocked out missiles. There were only two of these advanced fighters. Some missiles would get through.

Cirrus Minor Station:

The second destroyer was devastated by the second round of missiles once the fighters had left. They were launching troop shuttles, but only four escaped the blast from the destroyer's reactors imploding. The main plasma cannon and alexandrite lasers hit the third destroyer several times.

"Guild master, battleship has initiated a spiral turn and is now heading for the station."

"Target with the ion cannon. It is time to play our last card. Time for first troop shuttle?"

"We estimate 6 minutes. It is a living hell up there."

Mighty Mouse (Mining Sled Two):

"John, We are needed in the battle with the battleship, or many miners will die."

John immediately responded, "OK, Jemma, what are the survival odds?"

"Quite good, I would estimate eighty-five percent in favor of success. If we don't help, the station survival is twenty percent."

"Jemma, we are all in this to win," Sue claimed. The whole group nodded their heads in agreement.

Tony announced, "Immersion couches—helmets in the cradle by your chair. We will go full immersion. Same as before on guns." Then he sang a catchy tune, "*Mr. Trouble never hangs around when he hears this Mighty sound. Here I come to save the day. That means that Mighty Mouse is on the way! Here I am to save the day.*"

Jemma suggested to Tanya, "Railgun has three shots. Here is a schematic of this battleship. Hit these spots if we get a chance." The soft spots were highlighted on the hologram as red flashing points.

The sled accelerated once everyone was settled into their immersion chairs, which sealed them in and protected them from the tremendous g forces the sled was enduring. They rapidly closed the space between themselves and the battleship. Ten of the one hundred missiles launched by the mining sleds impacted the battleship's shields, causing them to go iridescent. While Chadoom's AI fighters were very effective, they missed killing seven missiles from the cruiser launch. The larger yield missiles caused the shields to go again iridescent and flicker, showing the massive overload. Then they came back, but with half the strength they had before. The little sled added to the pressure with a laser and plasma strike on a focused point on the shields, which caused the rear shield of the battleship to flicker again.

Isaacs enhanced fighter (T):

"Mackena, fire your missiles now. This is going to be hot. Two fast fighters chewed up the first wave."

"Acknowledged, Isaac, get out of there. Those fighters are heading your way."

"Launching anti-fighter missiles now, I am going to tear down the shield of this behemoth with a shot from the main cannon and Gaussian railguns in three, two, one."

The communication stopped as Isaac fired the main gun. It took most of the resources the fighter had. A sizeable green bolt ripped through space, striking the shields the mining sled had just harassed. The shield flickered again and died. Isaac immediately followed that with his giant slugs from his railgun.

As if in a magical ballet, the mining sled known as Mighty Mouse rotated onto the prominent fighter's flight path line, fired two purple beams, and added two more metallic slugs from its railguns. Then it rotated back, accelerating and flying evasively.

Isaac also fired two more slugs and remembered he had two enemy fighters on the way. The ion cannon from the station hit the battleship's forward shield, which flickered and died.

The station lasers ripped into the battleship, burning a hole straight through the front of the ship to a power junction that controlled the shields. All the battleship shields dropped in an instant. Isaac's first two heavy slugs ripped into the front of the battleship, shredding at least forty meters from the ship's front. The two smaller slugs had been targeting the bridge. One missed, but the other hit close to the bridge, severing it off the rest of the ship. Plasma cannons ripped into the large battleship turrets lining the vessel's side. The entire front of the battleship was now a flaming mass. The two AI had taken a liking to Isaac's fighter. They ignored the fifty inbound missiles to line up for a strafing run of what they figured to be.

Cirrus Minor Station:

The Guild Master was intensely watching the scene unraveling in front of him. "Report?"

"The third destroyer was incapacitated and drifting. Fifteen assault shuttles are inbound, targeting the western station arm. Battleship shields are down with mining sleds moving in from above—fifty missiles inbound in sixty seconds. Twenty-two Mantle Works fighters are still defending, plus the two AI fighters are now retargeting Isaacs's fighter – Trex. The rogue sled or Golden Sled is dancing with the Battleship."

"Very well, send Allison a request for them to surrender. She may ask them to surrender unconditionally. Target the shuttles to send them a strong message that we are not open for business!"

Battleship MF:

"Chadoom, what hit us? Is this as bad as I think?" Allison asked.

"Shields are down. Miner sleds with seriously modified plasma cannons are attacking from above, and I think an Ion Cannon on the base hit us with a shot. That modified fighter, I guess, is Isaacs. It and the mining sled are using mag-accelerated slugs at close range. No shields can stop those rocks." He said with a panicked look on his face. "What do you think? I say we button up and cut our losses."

"What?" Allison asked, with a growing realization she had lost.

Chadoom grabbed her by the shoulders, looking deeply into her watering eyes, "Get out of dodge and now! The bridge is gone, shields down with over ten inbound missiles. Time to go, my lady."

Chadoom punched the big yellow button.

Mighty Mouse (Mining Sled Two):

"Plasma cannons, one forward, two rear. These two AI fighters are monsters," Jemma commanded

Isaacs's fighter was taking a beating. His starboard shield was flickering. He had expended his anti-fighter missiles on the fighters who had been swatted down.

"We will conduct a strafing run. Folks, target lead fighter, make it count," Jemma said, encouraging his gunners, "Rear plasma gun target the second A.I. Fighter when it cuts behind us. It will try to knock out an engine."

Mighty Mouse twisted to line up on the first AI fighter, now pounding Isaac's fighter as he twisted and turned. The sled immediately fired with a short burst of the purple beam that leaped across, shortening distance immediately. The fighter's shield flared instantly. The bright green plasma cannon rocked the fighter as it slashed through the weakened shields and tore the back of the fighter apart. The plasma pulse cannon finished the job, leaving nothing but streaming debris.

The sled immediately turned at ninety degrees to loop back if the second fighter did not follow. The agile AI fighter did not miss a beat but had to dodge a plasma pulse. The erratic flight pattern of the two ships was like a well-orchestrated ballet, with the AI fighter maintaining a rear position, challenging the sled at every turn. It had three plasma cannons providing a continuous stream of deadly beams that must be avoided. The AI fighter could not line up a shot, but it was only a matter of time.

Isaacs enhanced fighter (T):

Isaac would also not quit. He had sustained damage and was struggling not to become a vaporized memory. He was falling behind—the torrid ballet. He smiled when he saw one of the Mantle Works AI fighters end in a massive fireball.

Mighty Mouse (Mining Sled Two):

"Jemma ideas?" asked a concerned Tanya as Isaac's fighter twisted, turned, changed velocity, and continued to evade the vampire on its tail.

"Working on it," Jemma said, stressed as she noticed an impact alert. "Warning, incoming object!"

Suddenly, the radio blared, "Tony, catch me on the flip side!" A silver streak crashed into the second AI fighter as Jemma dodged a piece of the battleship's nose. The fighter disappeared in a flash.

Tony immediately reacted, "That was Dad's sled!! Dad, Dad. We have to get Dad! That is a game we used to play on long trips called "rescue the miner." He jumped before crashing the sled. I know it."

John commanded, "Out of immersion, we have a Dad to find. Tanya, take over as a pilot. Sue sensors, Tony airlock suited. Jemma, please fly a figure eight, but keep evasive. I have weapons, but I am putting a hard shell on. Look for stragglers."

Sue coldly told the crew, "The Battleship will not be an issue in about four minutes. Missiles inbound, and we do not want to be here in three minutes and thirty-five seconds."

"Any other threats?" John asked.

"Lots of debris but no threats. Sleds are moving away homebound." Sue answered.

Tony was suited, "I'm in a hardcover. I have his trajectory, but we need to move now. Jemma, meet me in the airlock in thirty seconds. Grab two jet packs and a life cable."

Jemma woke out of her curled-up position and moved to her spacesuit. She was in instantly, and the suit was sealed. The suited dog stepped under a special hoist holding her jet pack. It clicked in place, and she padded to the airlock, meeting John. John checked her suit. She joined Tony in the airlock. They stepped out and fired up their thrusters, with Jemma leading. To John's amazement, they were targeting a small dot that grew into the form of a body in a soft mining suit, slowly spinning. There was movement, which thrilled Tony.

Jemma fired her suit jets toward their slowly turning target. When they reached Adrian, he hugged his son, Tony, tightly with big tears in his eyes. Tony tied an umbilical on Adrian's soft suit and tied him to Jemma. She would make sure he got back safely.

"Tanya, Come get us. I have my beacon on. We are maneuvering to you. Time is short."

"10-4. Be there in 30 seconds, Tony. Get in the sled fast. The missiles will hit the Battleship in one minute and 15 seconds. Would you please stay with the airlock and strap down? We need to move fast. The mining sleds have pounded the Battleship, and it is bleeding air and has fires everywhere. They are clearing space as fast as they can. Those miners know something; we better listen."

True to her word, Tanya pulled an emergency stop, spinning the rear airlock toward them twenty-one seconds later. The three boarded, and Adrian and Tony tied themselves against the wall of the airlock as it went through its cycle. Jemma dropped flat on the deck and wedged herself between John's legs.

Tanya asked, "Jemma if you need to drive, this would be a good time!"

Having monitored the situation, Jemma replied, "You're doing fine. I recommend you take it faster and, say, 90% power or better."

"10-4" Tanya boosted the sled. It was good everyone was tied down. The little ship jumped out of the Battlespace when Tanya hit ninety-five percent capacity.

Battleship MF. was now a burning sitting duck with fifteen additional missiles approaching. The mining base fired the ion cannon, which shattered the middle of the battleship, cutting a swath through the upper half. A flash blew out of the ship's bottom as a saucer-shaped disk flew out of the boat toward the asteroid belt. The missiles finished the battleship as they struck along the starboard side of the two-kilometer-long ship. The vessel became a small asteroid field - tiny scraps of metal and escape pods.

Cirrus Minor Station:

"All troop shuttles, we have a missile cruiser loaded for bear with anti-ship missiles. Surrender and move into a flat orbit. Power down weapons now. You will be escorted into the main shuttle landing bay one at a time." Mackena had moved the cruiser to block the assault suttles entry. She was mad as a mother bear with a cub. She had witnessed her baby, the fighter T, get abused.

"Your support is gone, and I have enough to feed each of your shuttles several five-megaton missiles. I am ambivalent about who gets it first, but my finger has a problem when it gets near this button." She moved her hand near the button and grabbed it with the other like she was in a struggle. "I just do not know if I can control this situation much longer. So General, how big are they? She smiled, taunting the leader of the shuttles.

"This is General Nashing on board Assault- shuttle 2. We acknowledge your request. We are standing down. Look, we were doing as we were told."

"Is the surrender unconditional?" asked the Grand Guild Master.

A short delay ensued. "Yes," came the reply.

The command center erupted in celebration.

Chapter 31
Cirrus Minor - Live For Another Day

Truth is better to tell early than to try to cover things up. Lies always unravel, and the truth comes to the surface. Truth is the essential element of success. Major news organizations in the 21st century coined the phrase "Fake News." Fake news is a bit of truth mixed with stretched, false, or misleading information presented as new. Major news organizations fighting for ratings would enter into providing misleading stories with one-sided reporting. These were joined by social media vehicles that would carefully stop anyone from posting an opinion opposite the company's leadership philosophy. Fortunately, that tool could only be used until the grand flowing stream of truth overruns it. The Truth is always out there – it is just how quickly you find it.

John met Tony and Adrian with a huge hug.

Adrian laughed loudly. "I figured it was you. I could not think of any sled on steroids other than you and your crazy dog."

"I think I should take offense," said Jemma, "but I don't have that emotion. Anyway, I also don't have the nickname Crazy Adrian!" She sniffed his hand.

They were now above the Trex, which had recoupled with its scorched fighter. Trex was now watching over the troop assault ships in orbit. Adrian finally looked around the mining sled. "What is all the stuff you have in your storage bins? The place looks like the inside of a hoarder's home."

"Dad, we are running heavy," Tony proudly said, "I think we found your Draconia Asteroid." John opened up one of the storage bins and pulled a large Draconia crystal out. Its iridescent flashes immediately lit up the room. Crystal thrall immediately captured Adrian, and stood silently mesmerized by the ever-changing iridescent flashes.

The silence was complete until Sue exclaimed. "Do I need to call the paramedics in here or what?"

They laughed. Then Tony said, "Wait until you meet our new friend Joel in the storage sled."

"You have someone in the storage sled, and you flew in a battle. That storage sled is as thin as paper and has limited radiation protection!"

"Mr. Joel was an old miner floating around your claim."

Adrian excitedly said, "A claim jumper? You should have spaced him. Now the mining guild will have to deal with him just as well."

"Well," Tony said, figuring he was going to have a little fun, "Turns out you and John's dad may have jumped his claim."

"Are you nuts! I checked, and Turk checked. We have the only valid claim," Adrian reacted. Claim jumping was considered an act of piracy and a capital offense.

"Mr. Joel was probably happy to see us. He had been hanging around that asteroid a long time."

Adrian, massively confused, said, "Are you pulling my leg? Is this April fool's day or something?" The youth laughed.

John rescued Adrian a little. "You and Dad cut away enough of the rock that you could extract the little vug for a prize."

Adrian, still puzzled, replied, "Correct."

"What you did not do after the sled was free was to inspect the rest of the cave. We did and found Mr. Joel, who had been waiting patiently for someone to pick him up."

"Why did he not bang on the front of the sled? I certainly did not hear anything, and no active claim has ever been recorded there."

"He could not knock because he is a mummy. He died hundreds or thousands of years ago." John explained.

Adrian, still confused, "But man has only been here a few hundred years."

"Guess what, Dad. Mr. Joel has a tail. An honest-to-goodness tail, which is why he was probably hanging around," the youth groaned at the pun.

"Are you telling me he is an alien?" Adrian exclaimed.

"We believe he is, and Jemma thinks his mother ship might be trapped on the edge of the keyhole. That was the clue you missed on your map. Who did you get your map from?"

"That makes sense. I picked the map up from a funny ole miner years ago. Quirky little fellow."

Sue said, "Mr. Joel is securely settled in the auxiliary sled within the quarantine zone we established. He's accompanied by his belongings, several large boxes of diamonds, and a small chest of draconia."

Jemma added, "He left a diary, which detailed his daily routines, intimate thoughts, and hopes and prayers for his wife, Quih, and their four children. I was able to cross-reference a map he left behind with Mr. Adrian's map and the asteroid database we compiled to begin deciphering the language. If interpreted correctly, all his children and pets had names starting with 'J,' which appears to be cultural."

"Wow, I thought I had run into some weird things. I hope you plan to get Mr. Isaac involved."

Jemma announced the incoming call. I will put it on the speaker."

"Sled 2616 to 2616b, over." Turk, the childreds father was calling, and he was mad.

Tanya turned white. Tony knew he would pay the price of adventure, and Adrian just smiled because he knew what was coming.

John answered," Mining sled 2616b, running heavy, over."

Turk was frantically losing his composure. "Reckless or just plain foolish, that was the most absurd piece of flying I've ever witnessed! John, you had your sister, your best friend, and Sue on board, yet you decided to play hero in the midst of a battle! You're older by just two minutes, but I swear I'll ground you for life if anyone is injured. What did you do to that sled, and who was flying?"

Tanya had had enough. "I was flying part of the time, and we all chose to participate, so ground us all, Dad."

"I will, but I need some help. Crazy Adrian destroyed his sled by ramming one of those advanced fighters. I think that wild-haired miner jumped ship before he destroyed his ship. I need to look for him. I need your help, and I hope that dog can smell him out."

Adrian, with a finger to mouth, hushed the young miners. "Adrian, to mining sled 2616, come in."

"Thank goodness, my good friend, where are you."

"Running heavy with these outstanding citizens, of which three are yours."

"You're a real devilish soul, Adrian, but I missed you."

"Same to you, buddy, let's meet on Mr. Isaac's ship. He needs to know we are heavy and have a special guest quarantined in the utility sled. The kids need to explain this one."

"10-4 sled 2616 out over".

"Mr. Isaac was listening on the line," Jemma announced.

"Put him on," Adrian said.

"Thanks for the save, Sled 2616b," Isaac said. "Can we give you a lift to your new home?"

"Yes, I don't have a sled anymore. Can I bring some stray kids along?" Adrian asked.

"Why, Adrian, you must be the luckiest man in the galaxy. First, that vug, and then you pull a crazy stunt of taking out that AI fighter and surviving, Hoorah! Bring the whole herd."

"Mr. Isaac," John stepped into the conversation, "We are running heavy, and we have a quarantine situation in our utility sled. All is contained, but we need a quarantine area in your ship bay."

"Not a problem. We have an area we can isolate. Do I need to get a medical team to the ship?"

"No, sir, we will explain when we get there?"

Then Isaac asked, "Is Mister Golden still on board? I want to meet him."

"Only his spirit and handy work remain on board. Mr. Golden dissipated," John replied

"Dissipated?"

"Yes, you know he completed his work and vanished."

"You will need to explain that also. Very well, hurry. I need to beg Mackena's forgiveness for getting in over my head. The sooner you get here, the less I must plead." Isaac signed off.

Chapter 32
Exiting Cirrus Minor - Escape for Another Day- Near Cirrus Major

Evil manifests itself in many ways. It is sad to say evil creates evil. Some call that the proof of the Chaos Theory. The only way to change that is to change from the inside. Some do not recognize how difficult that is.

"Chadoom, are we on course to Cirrus Major?"

"Yes, Ma'am, I have called our base, and our destroyer will meet us halfway. They pulled out of the repair dock, and the repairs were halfway complete. We will return it to the dock for another two months."

Allison asked, "It's been there for three months already. Did they run into an asteroid around Earth?"

"Well, they accidentally ran into an Earth force destroyer, cutting it in two. Quite unfortunate for them."

"Did they do it through carelessness?"

Chadoom replied, "No, remember that Earth Force Captain who was so vocal about our operations in the Earth's asteroid belt."

"Yes, I think I wished him dead."

Chadoom looked at Allison and calmly said, "Your wish is my command."

"Oh, did you have to wreck a destroyer?"

"Part of the job, mam," Chadoom replied matter of factly.

"Sensors are detecting two ships ahead of us. Sensors indicate these are damaged ships. We have a bonus. They are our two damaged fighters, sending friend or foe ID codes. We will take them under our wing, so to speak."

"These are artificial intelligence, right?" Allison asked.

"Yes"

Allison offered, "Well, let's fix them at the dry dock and use them to build an army. We will create our complex on one of the planets outboard of Cirrus Major."

"With their most recent experience and your drive, that complex could produce an army of mighty ships. I will shift the funds remotely. You start the pre-feed on the engineering. There will be some significant fallout with the board associated with this disaster. If everything goes wrong, I will move funds into our accounts to cover this operation."

Chapter 33
Cirrus Minor - Victory Dinner - high orbit Trex

To the victor goes the spoils. What good and light will come from that victory now that you have won?

Isaac and the guild master met the two sleds as they landed. They moved the utility sled to an isolated part of the bay, temporarily cordoned off under a negative pressure tent. The group gathered at the base of Sled Two, now called Mighty Mouse.

The guild master beamed with satisfaction. "We've sent the fifteen troopships back to Cirrus Major to reunite with their families. The miners decided to let a group of down-on-their-luck miners handle the salvage from the battle. They'll receive sixty percent of the proceeds, with the rest shared among the miners who fought. The Mantle Works Cruiser wasn't destroyed and is yours to salvage as a special thanks for providing a solution. Mining engineers who boarded it say the Cruiser's evisceration is total, and it should simply be scrapped."

John spoke. "We were only doing what that old miner asked us to do. You know, the one with the grabblely voice and the funny laugh. Mr. Golden is resourceful and scary at the same time if you ask me. We don't need the cruiser. Would you please give it to the others? Everyone in agreement?"

Isaac brought the attention back to the mysterious Mr. Golden. "I still want to meet that old wizard."

"Well, he did equip the base with advanced new weapons." The guild master added.

"When do I get to meet him in person," Isaac asked

"I have not even met him except via video link," the guild master explained.

"A real recluse, and he said we had already met him - I cannot think of when or where."

Jemma walked out of the ship, tail wagging. She carefully greeted Isaac with a good sniffing, sat down before the guild master, and extended her paw to be shaken. Isaac noticed the dog was sporting a new gold collar with a small shield-shaped tag with the letters CTR stamped on it. She had been around Mr. Golden, he thought, stumped for a moment.

"Such a good dog," the guild master commented.

A hologram flickered to life beside the group. The elderly miner, Mr. Golden, stood there with his shaggy dog by his right side. "My name is J. D. Golden," he said, pausing to adjust his robe, which rustled softly in the dark. Isaac noticed a pendant glinting on a delicate gold chain with the letters "CTR." Golden continued, "I'll be close by if you need me. For now, I'm off to assist someone else in need. Well done, my friends."

"But we need you." the guild master said.

"When you need me, I will be there. When you want me, you won't see me." And the image disappeared.

"Wow, that was interesting," said Isaac.

"Sir, sled 2616 on approach vector." A crewman announced.

"Good, I am hungry, and Turk is delaying my lunch," Isaac stated, shaking off the unexpected visit.

"Mr. Isaac, we have some interesting data we need to show you related to our quarantine. We need your advice on how to proceed."

"Well, it sounds like lunch entertainment. Let's collect your dad, John. I have a pickle waiting."

"What a pickle? I have not had a pickle in years," Adrian exclaimed.

"Dill?" the young man asked.

"Is there any other kind? Isaac firmly stated.

Turk landed his shuttle. There were severe burn marks from laser strikes and at least one skin penetration. He was unscathed and ran out to hug all four children. Even Jemma got a hug and understood that dogs in this type of situation should provide a lick on the cheek, which she masterfully executed. "I thought about it, and while I am mad as heck with you kids, I am extremely proud of you."

Tony added more fuel to the fire."Wait until he meets Mr. Joel!"

"Who?"

John jumped in before Tony could light a match as he had done with his father. "Dad, we will explain after dinner."

Isaac announced, "Pickles are waiting. Oh, what do you feed your dog?"

Tanya, thinking fast, said, "She was fed earlier, and John, no table scraps, or she will get fat."

Sue added, "She does not look like it, but she is a high-maintenance dog who is on a high-energy diet. Tough to handle on a small mining sled."

"We can put her on Mackena's treadmill if you need," Isaac offered, leading the way.

Adrian, walking beside his lifelong friend, "Why is it that your kids wipe out a cruiser, take out the front of a battleship, and knock out an advanced fighter, and I lose my sled, and you look like Swiss cheese? Do you see any battle marks on their sled?"

"I think I saw a small scorch mark on the starboard side of the sled, and if you don't see it too, I will go put one on there. Got it?"

"Yeah, pretty amazing, and I am sure I saw the same mark."

Jemma slapped her tail hard against Adrian's leg.

"Ouch, smart ass," Adrian exclaimed.

Sure enough, the table was a celebration banquet. And as promised, dill kosher pickles were available in abundance. Tanya gave a synopsis of their mining track. Tony filled in the group on the keyhole. John described the exploration into the cave where they found the mummy they had named Mr. Joel. They just assumed he was a male. "Another space-faring race was mining these asteroids long before we arrived. In some ways, they were advanced to us; in other ways, they were primitive. She told them of the diary and its contents, the chests full of diamonds and draconia.

"What is the biggest clue of the aliens' origin?" Isaac queried.

"Simple," she switched the halo to a graphic that showed Mr. Joel's tail. "We also found a map that, with computer effort, we could trace objects that had moved relative to one another. We especially rectified them with respect to their movement, which helped us with another problem that we won't describe until we have an iron-clad confidentiality contract."

Her wrist computer beeped. She read the short message, "I understand you have signed."

"I did indeed." Isaac smiled.

"Good, we will not show it tonight, but we have a working model with a proven track record of the destruction of Cirrus Minor."

Sue approached the holographic projector and changed the picture to the rock face loaded with Draconia.

"Wow" was the only word said for a few minutes as the projection moved around the face.

Tanya brought a box in that Jemma had fetched. Miss Mackena, we understand you like interior decoration. We hope this will be a proud addition to your Christmas display." She gave the box to Mackena, who smiled kindly and opened it. The dazzling radiance immediately radiated through the room but mixed with a ruby-red flash.

"I cannot accept this. It is your future." Mackena stated.

Tanya, ready for rejection, countered, "You are honest, truthful, and good to your word. We have a new home and a bright future because of you."

Sue added, "We have two other special boxes for the moms. I am fortunate to find a loving family after my parents' death. We thought we should celebrate the better ladies behind good men. This gift is from our hearts. Please accept it as intended."

"Well, putting it that way, I cannot refuse." Mackena stepped toward the young ladies and hugged each of them.

Isaac moved the subject back to the alien. "The alien and its artifacts are a puzzling problem. This is a significant find and has severe ramifications for the human race. Yes, you prove that we have not been alone in the galaxy and intelligence exists elsewhere. We have always maintained that.

Interestingly, this alien exhibits many human-like qualities, with the addition of a tail being particularly intriguing. It's likely that some, including governments, will perceive it as a hoax. Aliens tend to be problematic for governments. The artifacts and language are intricate and clearly not of Earth, and further analysis of the DNA, notebook, and map composition will offer crucial insights. It's essential to share the full story with a comprehensive analysis and encourage others to study it. If the alien is turned over to a government, they'll likely either bury the evidence or, worse, attempt to weaponize it. To prevent it from vanishing, a neutral research base is necessary.

You've taken the right steps by preserving everything and quarantining the creature and its belongings, avoiding potential exposure to something hazardous. Many institutions would be eager for the opportunity to research this. But is humanity ready to accept proof that we're not alone? I believe so; after all, science fiction entertainment has prepared the public for this reality."

Isaac stopped and thought a bit, then continued. "I suggest we set up our research center on an isolated asteroid and bring in some of the most talented people in various fields as needed. We will start with the special construction of an isolated special containment unit to understand the infectious risks we are dealing with. It will be expensive, but I can see this multifaceted research that could lead to substantial breakthroughs. The question is, do you want to participate? You can, by the looks of things, here. And I am guessing another mining trip, and you will own the Draconia markets like the old Debeer's of the Earth did diamond. You will need to manage that carefully. I would love to help you. I need another really good project."

Sue quickly asked, "What would your management fee be?"

"I think one credit a year plus 5% of the net profits, and Mackena might be interested in helping with your new Alien Studies and Exploration Institute."

"Hmm, that sounds attractive. Does that make you a board member?"

"Yes, as Chief Operating Officer."

Tanya added, "This sounds good to me, but I must insist that Mackena receive a board position at the institute. We need someone we can trust who also likes dogs."

Jemma had spent the evening with her head resting on Mackena's lap, seeking a scratch, and John thought she was one of the biggest attention-seekers he'd ever seen. Meanwhile, Twitch kept circling behind Mackena's chair, either guarding or perhaps just very jealous. Eventually, Jemma approached Twitch, and they touched noses. It seemed like Jemma was hacking into Twitch's programming to convey that everything was fine. Once the nose-sniffing was done, Jemma returned to her "give me a scratch" routine while Twitch sat down calmly.

"Now I hear you have another crystalline vug," Isaac asked.

"Yes, we do. It is bigger than the previous vug and has not been opened. We also have a smaller one. The large vug has been scanned." Tony opened another holograph file, like a 3D cat scan. Like the first, the inside was spectacular and contained a couple of Draconia crystals.

"This belongs to our new mining corporation. Since you will be the CFO, this will help solidify the company's financial basis. We are happy to provide a 7% overriding royalty."

Isaac immediately retorted, "I believe I indicated 5%, but beggars can not be choosers."

"We considered 5%, but when we discovered that you made 200 million credits on the last vug, we realized your contacts would be worth the extra." Sue indicated.

"Ok, I will, except, but how did you find out what I sold the first vug for?" Isaac was puzzled.

"That is our most interesting resource, and we will share that knowledge when the time is right," John indicated, and he continued. "Uncle Adrian and Dad, we had not discussed this with you, but we would like your agreement."

Adrian's eyes lit up, "Kids, I am only a simple miner, and I am happy to remain that way. The head of a large corporation dealing in draconia, diamonds, and aliens scares me. I agree this is a fair offer."

Turk was now where the attention was focused. He smiled. "Isaac, this is like inheriting another brother. These kids have more on the ball than I ever suspected. Heck, I am kind of like Adrian. I am really happy mining, but I can add value to our company differently than mining. I agree with this proposal."

"Dad, we thought that through too. We were hoping you would be the company's president. And Uncle Adrian, we would like you to be a Vice President in charge of mining."

"As a chief operating officer, President, and Vice President in good hands, we will have time to attend university and return to specific research areas. We have a list of suggestions we will share with you later, as well as our understanding of the prospectivity of the asteroid belt, which I think you will find very interesting."

Sue continued, "I have a new family. We have been through an incredible adventure in a short period of time that has solidified the bonds of companionship. Thank you"

Turk hugged her. "You are the other daughter I always wanted - welcome. Families are forever."

Jemma rolled her eyes and put her head back in Mackena's lap for another ear scratch. What an actress!

Other Books by this same author:

Encounter with the Hemogeny Books 1-6

(2024-26 release).

Book 1: First Contact Completed: release target 2024

Captain Henry Teach, leading a crew of scientists and unconventional officers aboard the freighter Curious and its tiny escort, sets out on humanity's first interstellar contact mission. Their message of peace is met with violence, forcing them to adapt to a harsh reality – Humans are either prey or free labor for the dominant alien species, the Hemogeny. With their escort frigate destroyed, Teach must devise a way to escape without revealing navigation secrets. Using their ingenuity, the crew of the Curious builds unconventional weapons and captures an alien pilot, a desperate gamble that changes the fate of humankind. Join Captain soon-to-be Admiral Henry Teach as he creates a new future and changes human destiny.

Book 2: End of the Pirate Oligarchy Completed

Book 3: Freedom for the Oppressed Completed

Book 4: The Shattered Realm Completed

Book 5: Backdoor (In progress) Release Target 2025

Book 6: Paradigm Shift: Are We the Precursors? (Storyboard) Release Target 2026

My Way or the Highway release date 2024

In 2030, an honest, respected, retired geologist becomes the victim of an out-of-control, politically appointed, corrupt DC US attorney. This book has its roots in the Jan.6, 2020, riot at the white house, where later the U.S. President promoted the prosecutor leading the defunct January 6th Congressional committee to the coveted D.C. Attorney position. This shady prosecutor, however, has a black book that he levers to uncork the power of the U.S. government to satisfy his thirsty, corrupted ego.

The retired geologist, John Tsavorite, has settled into the local community and owns a small property outside Star Valley, Wyoming. He now works with his fleet of drones and a dog named Jadda while serving as a prominent volunteer in the small mountain valley town.

With no tangible information, the D.C. Attorney identifies John as an insurgent terrorist leader, Waldo, who organized the January 6th insurrection and had never been identified. The prosecutor uses his office and other government resources to investigate and capture Waldo. Many of the government resources result from the prosecutor using his black book.

Follow the clever inventor John Tsavorite, who, with the help of trusted local, honest people, fights the tsunami wave the U.S. Government cast down on the small, once pleasant town of Star Valley, Wyoming.

TIC TAC Status: Writing Phase (Release 2025)

TIC TAC is a book set in the very near future. Foo Fighters, TicTacs, Saucers, Spheres, Jellyfish, and the dreaded Triangles mixed with a few Global Leaders, Navies, Armies, and lots of fissionable material are making this story Very HOT!!!!!!

The Author:

Joel James Guttormsen is a father, husband, gemologist, and geologist. Joel has traveled the world, living in several countries for over sixty-five years. Joel is a scientist first and a writer second, with over forty years of using geoscience in the quest for significant discoveries. These discoveries include food, dogs, people, gemstones, and oil and gas. Joel is also a gemologist and a professional gemstone cutter. Joel has four children and a loving wife who will all testify they have been exposed to geology, but no one has said - enough.

Joel also loves dogs, which play essential positions in his writing. These furry friends manifest themselves bigger than life as Joel explores AI in the lives of his characters in the novels he writes.

Joel has never been accused of lacking the imagination that needed to come out (so to speak) stimulated by the long rides in and out of London where this book was written. He lives in the United States and is surrounded by the things he loves. A warm house, a dog curled up almost on his feet, the radiated energy from complex mineral compositions surrounding him in his office, and a blank sheet of electrical paper. Have you ever imagined how to twist a small rock like a folded piece of paper – I have one.

Please Imagine and Enjoy

Printed in Great Britain
by Amazon